Praise

"*Space Wars* is a book tha[...]
to arms, or as a fictionalized exploration of the world of wargaming and space security. In any sense, though, it's an original."

—*Defense Technology International*

"Unless we are willing to accept the fact that there are some out there who are fanatically bent on destroying us, *Space Wars* could be profoundly prophetic."

—Capt. Eugene Cernan, U.S. Navy (ret.), pilot
for *Gemini 9, Apollo 10,* and commander of *Apollo 17*

"The hair on your neck will rise as you absorb the latent danger facing the United States as revealed in *Space Wars*. The utterly believable, totally logical premise is terrifying, and the book should be made required reading by Congress. As *Space Wars* makes clear, we are on the edge of a low-tech catastrophe that will wipe out our high-tech advantages."

—Col. Walter J. Boyne, U.S. Air Force (ret.),
New York Times bestselling author of *Supersonic Thunder*

"*Space Wars* is a realistic, often startling behind-the-scenes account of military space strategies and the mounting geopolitical threats in our world today. It will leave you gazing toward space with a new understanding of the great battlefield just beyond our hemisphere and the critical role it plays in the future of mankind."

—Lt. Governor Jane Norton, State of Colorado,
delegate to the Aerospace States Association

SPACE WARS

THE FIRST SIX HOURS OF
WORLD WAR III

William B. Scott

Michael J. Coumatos

William J. Birnes

FORGE®

A TOM DOHERTY ASSOCIATES BOOK
NEW YORK

Although the characters and actual events depicted in *Space Wars* are fiction and bear no relationship to actual persons living or dead, the action and weapons depicted herein are based on actual technologies and war gaming strategies used by the United States military and civilian war planners in preparation for the types of events depicted fictitiously in *Space Wars*.

SPACE WARS

Copyright © 2007 by William B. Scott, Michael J. Coumatos, and William J. Birnes

A Forge Book
Published by Tom Doherty Associates, LLC
175 Fifth Avenue
New York, NY 10010

www.tor-forge.com

Forge® is a registered trademark of Tom Doherty Associates, LLC.

ISBN-13: 978-0-7653-5900-1
ISBN-10: 0-7653-5900-6

First Edition: April 2007
First Mass Market Edition: October 2008

Printed in the United States of America

0 9 8 7 6 5 4 3 2 1

Dedicated to Linda, Erik, and Kevin Scott,
who sacrificed the weekend and family-vacation time necessary to write this book.

And to Nancy Hayfield Birnes,
whose support got me through four years of law school,
the California Bar Exam, and the long march
to complete this manuscript.

And dedicated to the American fighting man
and woman—standing lonely vigils across the globe . . .
waiting.

ACKNOWLEDGMENTS

The authors acknowledge our editors, Bob Gleason and Eric Raab, for their kindness and forbearance through multiple drafts of this manuscript and our publisher, Tom Doherty, for his courage in publishing a work about a war in space. Thanks also to Gemini and Apollo astronaut Captain Gene Cernan, who provided invaluable explanations of orbital mechanics; Major General (Ret.) Doug Pearson for his suggestions regarding antisatellite attacks from an F-15 fighter; Generals Joe Ashy, Howell Estes, Dick Myers, and Ed Eberhart, all former commanders-in-chief of U.S. and Air Force Space Commands, for sharing their concerns about the fragility of today's U.S. space infrastructure; Drs. Dick Wenham and John Kleiner for keeping me in the game; and Andy Cameron for his "hole-in-the-sky" idea. Heartfelt "thank-yous" go to Tony Velocci, Jim Asker, Dave Hughes, and Dave North, key editors and managers of *Aviation Week & Space Technology,* who provided us the equivalent of a graduate degree by assigning him to cover military space issues for the magazine—and paid him to do so.

Thanks also to George Noory for his foreword and for the many people, who must go unnamed, who provided us with the insight necessary to complete the manuscript.

FOREWORD

by George Noory

Space Wars is a distillation of outcomes from several space-oriented wargames conducted by the United States military services in the late 1990s and early 2000s. By mixing those outcomes with the oft-stated concerns of four-star generals in charge of protecting U.S. space assets, authors William B. Scott, Michael J. Coumatos, and William J. Birnes created a hypothetical "DEADSATS" wargame that addresses real-world scenarios that could trigger the beginning of World War III. Although "DEADSATS" is called a "wargame," it is anything but a "game."

The events depicted in *Space Wars* meld lessons learned from actual military wargames, such as the Army-After-Next, Navy Global and Air Force Global Engagement series, Space Game 2 and Schriever 1 & 2, with real-world geopolitical events and forces. These and other wargames are part of national security protocols that examine a range of worst-case scenarios—including electronic, kinetic, and cyber attacks on American commercial and military space platforms.

Space Wars depicts strategies and tactics that constitute modern wargames, but takes them a step further, to show how wargaming can be a valuable real-time decision-making aid for national leaders. Wargames may be conducted deep within our nation's most secure military facilities, but their outcomes rarely influence national policymaking today. This changes in the 2010 setting of *Space Wars*. Classified wargames are being conducted today, testing the military's various options for responding to threats, and exploring the ramifications of preemptively disabling those threats, for example. But they also amalgamate military, political, and diplomatic solutions into a global strategy for "waging peace, not war," as President Dwight D. Eisenhower said.

Author Michael J. Coumatos (U.S. Navy, Retired), a naval task force commodore and combat commander during the first Gulf War, was instrumental in developing wargame strategies for United States Space Command and, later, for commercial companies. His experience in designing, developing, and managing wargames underpins the reality of this book. But in *Space Wars*, authors Mike Coumatos and Bill Scott, retired Rocky Mountain Bureau Chief for *Aviation Week & Space Technology* magazine and a former U.S. Air Force Flight Test Engineer, not only depict how actual wargames can contribute to the understanding of future threats and conflicts, they also tightly link wargaming to realistic scenarios that could very well become headlines in the next few years.

The authors' graphic descriptions of just how vulnerable the world's space infrastructure really is expose a critical national Achilles' heel that politicians, economists, and corporate CEOs have largely ignored. Precisely how dependent upon space the Western world, in particular, has become is illuminated from a perspective rarely examined in a public forum: What happens when an enemy attacks and quickly disables military, intelligence, and commercial spacecraft. Such losses in space can quickly affect the economic, social, and national security fabric not only of the United States, but of the entire world.

The book explores how rapidly economies can be disrupted, as if molasses were poured over an entire society; how a handful of clever bad actors can embolden rogue nations to take advantage of a country's degraded abilities to watch, listen, and communicate; and how large military powers can be held hostage by the unknowns inherent in a new kind of war.

If America's leaders could stand on a mountaintop in 2010 and look back along history's path, what road would they choose in relation to the so-called freedom of space? An appropriate predecessor to consider would be the United

States' history as a seafaring nation. Her control of the seas has ensured stable free trade during peacetime, and has guaranteed the transportation of men and materiel to far-flung corners of the world when the world convulses at war. American sea power reinforced U.S. allies in World War I, defended convoys against the Nazi U-boat onslaught, and swept the Japanese navy from the Pacific in World War II. When Kuwait was invaded in 1990, America's navy became the backbone of Operation Desert Shield, quickly moving tons of equipment and combat manpower to the Persian Gulf.

But in the twenty-first century, while the concept of "freedom of the seas" is well established, the free use of space is being challenged. Will the United States ensure it succeeds as a spacefaring nation as well as it has as a seafaring power?

America's dependency upon space-based systems today makes freedom of space every bit as vital as its sea equivalent. Now, the transport of timely space-derived intelligence data is through communications satellites, and those data, turned into knowledge, are supporting critical decisions. Any disruptions of that fragile space infrastructure have immediate, serious consequences.

This space dependency—and the associated space infrastructure's vulnerability—are addressed by *Space Wars'* "DEAD-SATS" wargame, set in 2010. However, the wargame quickly becomes entangled with reality, when space leaders draw on game participants to assess a number of spacecraft anomalies. Their mission: to determine whether U.S. satellites are dying from natural causes, such as cosmic radiation and micrometeoroid impacts, or are under attack by an unknown enemy.

In *Space Wars: The First Six Hours of World War III* the authors have given the wargamers and protagonists fictional names, companies, and locations, but the threats are very real. The story depicts events that very well could unfold, given the vulnerability of today's spacecraft and the volatility of both rogue nations and non-state actors, such as terrorist groups and drug cartels.

The technology and weapons featured in the book are not science fiction. They exist. While still highly secret—and unacknowledged by the Pentagon—the "Blackstar" two-stage-to-orbit spaceplane system is real. The successful shoot-down of a satellite in orbit by a missile fired from an F-15 fighter is real. And the capability of Iran to launch a nuclear-armed intercontinental Shahab ballistic missile is all too real. In fact, Iranians were bragging about that capability as the authors were writing this book.

Mike Coumatos and Bill Scott take us into a frightening, very real world, where narcoterrorists control huge amounts of money that enable buying and deploying their own warfighters. The Cali and Medellin cartels, for example, once claimed financial resources greater than those of their host country, Colombia. Narcowarlords owned telephone and cellular phone companies, allowing criminals to monitor the communications of paramilitary agencies that were tracking them. Like American organized-crime figures of the 1930s, who controlled local governments during the Prohibition era, narcowarlords also control cities, governments, and armies. What happens when narcoterrorists team up with political terrorists? Unfortunately, that's happening—and is one of several driving scenarios in *Space Wars*.

Coumatos and Scott go beyond today's Global War on Terror headlines, taking readers into a disturbing, far-too-possible future—a place where America is reeling from unrelenting onslaughts that threaten to destroy the fabric of its society. Their story begins with strategic upheavals set in motion when the United States and its allies retreat from a decade of gains throughout the Middle East, following the 2008 presidential election.

There is great concern among America's allies, trading partners, and emerging Third World nations that history's next cataclysmic upheaval of empire is at hand. Unrecognized by all but those who see the big picture, critical clashes have already begun. For instance, few of the world's intellectuals, statesmen, and military leaders understand the subtle implications of a few satellites simply going silent in the cold blackness of space. In reality, those are the first shots fired in a clash of civilizations. Their echoes will grow to tectonic proportions, with terrorist attacks just one of several horrors employed by radicals committed to reshaping the world into the model of seventh-century Islamic caliphates.

For radical jihadists imbued with fundamentalist Islamic tenets, Charles Martel's defeat of Islam at the Battle of Tours in 732 happened only yesterday, the fall of Jerusalem to the Crusaders only this morning, and the 1948 partition of Transjordan and the land of Israel only a few minutes ago. But tomorrow is reserved for revenge.

And there are other dangers. An amalgam of nation-states and netherworld bad actors pledged to the obliteration of America's influence and wealth are quick to recognize opportunities presented by the degradation of space-based communications, navigation, and surveillance capabilities. Several of these entities rapidly mobilize multipronged efforts aimed at exacerbating and fueling the spreading global chaos. *Space Wars* may be a set of hypotheticals, but its scenarios are only extrapolations of space-related attacks that have already been made on the Global Positioning System and several commercial satellite downlink signals in recent years. These dangers are very real, and growing.

What does this all mean to America and the free world? Moreover, will decision makers recognize these shifting forces and apply lessons from the past, from the experience of space-smart leaders, and from wargaming's proven capabilities for conducting "what-if" analyses, to explore various courses of action and to predict an adversary's intentions across an array of scenarios?

Space Wars highlights the need for new approaches to classic thinking, when it comes to the still emerging frontier of space. America's best and brightest must develop and apply dynamic decision models that consider not only the technologies associated with space, but also the evolving nature of the human condition as it relates thereto.

Space Wars suggests that wargaming can provide political and military leaders a dynamic means for looking across a range of futures to assess various political and military courses of action. It also could enable vastly improved decision cycles, allowing leaders to consider multiple potential outcomes, accompanied by intended and unintended consequences. The most likely outcomes, those that yield strategic and operational options, could quickly be incorporated into political and commercial strategies, as well as traditional military war planning.

Space Wars showcases the potential benefits of assembling a wargame group comprising defense laboratories, the private sector, academia, and various government agencies outside the intelligence and defense communities. Capitalizing on the diverse, collective experience and skills of scientists, mathematicians, statesmen, corporate executives, and military leaders, the DEADSATS wargame task force embarks on a crash program to develop a simulation-based system-of-systems based on complexity theory. In the process of helping defuse a global crisis, they become the founders of a new methodology for decision making, aided by advanced technologies and knowledge-management tools.

The ancient art of wargaming was described by Sun Tzu in his *Art of War*, a methodology employed by Chinese emperors and Japanese warlords, and practiced by Carl von Clausewitz. To be clear, today's wargaming has been adapted from the multiple facets of traditional gaming as understood by earlier practitioners. This included assessing political and military decisions while gaining knowledge of strategic and operational outcomes—all while incorporating adversarial and nonaligned elements into the process.

However, these adversarial and nonaligned representations were sorely lacking in most approaches to decision making at the highest levels of U.S. government. The missing factors could be found in military wargames, though, wherein Red (the threat); Green (non-aligned, commercial interests, media); and Gold (allies) teams were institutionalized.

As a premise for *Space Wars*, a cadre of wargaming expertise has been built up within the uniformed services over the years, representing the real world of many conflicting interests. But such confrontational gaming methods have not been readily adopted by the U.S. intelligence community (particularly the CIA), FBI, and Department of Homeland Security, nor have wargame protocols been embraced for National Security Council (NSC) deliberations. Every agency still looks at world events from within its own silo, with little or no communication from the outside. Ideas are recirculated and recast in different terms, but there is no paradigmatic challenge to corporate mindsets. Thus, no new ideas can get into the mix.

As the pace of geopolitical problems and associated military consequences increases, military planners must recognize that the introduction of a more rigorous decision-making capability is crucial. While technological solutions are vigorously pursued by the larger community, a wargaming cadre within a U.S. Strategic Command task force embarks on a path of what can best be described as "learning"—both institutional and individual learning—to evaluate multifaceted elements of strategic thinking and decision making; learning how to counter hubris through introspection; and learning the art of seeing problems, actions, and consequences through the lens of an adversary, an ally, and a fence-sitter.

The wargaming cadre had studied history ranging from the German *Kriegspiel* (using scale models from the early nineteenth century) to the singular genius of John Boyd's *Observe, Orient, Decide, Act* (OODA) *Loop*. Physicists and engineers were introduced to wargaming fundamentals grounded in sociology, anthropology, and philosophy. The institutional representatives of a growing wargame community learn and relearn, and thereby build their own bedrock of principles that augment developing technologies.

Wargamers ultimately adapt their methodologies to align with NSC protocols, while complementing processes already well established in the Defense Department, the uniformed services, and the worldwide network of U.S. military combatant commanders. Several key political decisions have been derived through wargaming, and the favored tool of military leaders has found its way into the culture of top-level governmental process.

The authors apply this new decision-making paradigm to the

Space Wars world of 2010, when rapidly unraveling geopolitical situations challenge America's leaders. The clouds of war are gathering. The presidential election of 2008 has vastly shifted America's approach to world engagement. The new president exhibits uncertainty in the wisdom of America's original strategies in dealing with terrorism and radical Islam. American citizens suddenly experience suicide bombings that once were the deadly province of only the Middle East.

By early 2010, a highly advanced knowledge-based computer system also is in beta-testing deep within a classified area of U.S. Strategic Command at Offutt Air Force Base, near Omaha, Nebraska. Called BOYD, this secret system, much like the original National Security Agency parallel-processing computers, far exceeds the technology levels of mainstream, unclassified computers.

In an early test run of the system, aimed at highlighting allied vulnerabilities, several aspects emerged, reinforcing the more subjective outcomes of late-1990s, early-2000s wargames. Most alarming was the identification of improved, effective relationships among Islamist groups and their ties to rogue nation-states, drug trafficking cartels, and international criminal consortiums. In effect, the U.S. and its allies now face a web of nontraditional, shadowy enemies that attack in vicious, unexpected ways.

The BOYD prototype runs are frightening. Threats are evolving from suicide bombs and clumsy, but menacing, attempts at employing weapons of mass destruction. These threats now unleash a new wave of warfare, including attacks on space platforms. Simulation after simulation show the time-compression and cascading nature of such attacks, where events literally spin out of control, surpassing what Malcolm Gladwell has called a "tipping point." There is little to no opportunity to assess the problem, analyze options, make decisions and take actions. BOYD says America could be in mortal danger. Then events in the real world start tracing BOYD's and the DEADSATS wargame's ominous paths. And America remains woefully unprepared.

This book is an account of unfolding events in the spring

of 2010, triggered when a satellite orbiting high above the Earth simply dies. Then another goes silent. And another.

Lady Margaret Thatcher, the former prime minister of the United Kingdom, once said the following about the world situation:

> Of course, the identity of the source of the threat changes. Yesterday, we could have said with reasonable confidence that at its root would be some degree of Soviet mischief-making. But today, Islamic extremists, ethnically driven terrorist groups, rogue states no longer disciplined by powerful patrons—all of these have assumed a new importance, alongside the age-old problem of the dictator in charge of an unstable, bankrupt, expansionist state . . .

Space Wars depicts an uncertain, chilling world of tomorrow, yet it was aptly framed by Thatcher's premise almost ten years ago. Today, Islamist fervor for terror is escalating, and rogue states are flexing their muscles in a cauldron of geopolitical chaos. And as there have always been, opportunists abound.

In 2010, the United States could easily find itself in a national security dilemma as both commercial and military satellites mysteriously go silent. Perhaps other free-world countries experience the same type of dead-satellite trauma. What is the cause of such orbital fatalities? Who is to blame? What can be done about them?

Are such scenarios simply science fiction? Before you decide, consider the *Space Wars* story. Fiction? Yes, but it's also as real as tomorrow's sunrise—albeit a sunrise in 2010.

LIST OF CHARACTERS

Pierce Rutledge Boyer, President of the United States (POTUS)

T. J. Hurlburt, U.S. Secretary of Defense; retired Army general

General Howard Aster, USAF; Commander, U.S. Strategic Command (STRATCOM)

J. D. Hart, NASA and Department of Defense (DOD) troubleshooter

Colonel Jim Androsin, USA; Chief, STRATCOM Wargaming Center

Jack Molinero, Vice President, TransAmSat, a satellite company

Colonel Adrian "Matt" Dillon, USA; Commander, Army Space Command (Forward)

Senator C. I. Creighton, U.S. Senator, California

Lieutenant Colonel Thad "Burner" Burns, USAF; aide to General Aster, STRATCOM commander, and former B-2 bomber pilot

Annie, General Aster's executive assistant

Brigadier General Hank "Speed" Griffin, USAF; test pilot and USAF military liaison to U.S. National Security Council

Lieutenant General Dave Forester, USA; STRATCOM Director of Operations

Rob Joaquin, Pantera Corp. representative and satellite engineer

Staff Sergeant Sean Cantrell, USA; leader of Special Operations Forces team in Algeria

Captains "Pepper" Malloy and "Shark" Fisher, USAF; F-22 pilots over Algeria

Staff Sergeant Mitch Kucera, USAF; satellite operator, USAF 2nd Space Operations Squadron (2SOPS)

Martin Timm, National Reconnaissance Office (NRO) Director

Jill Bock, STRATCOM Wargaming Center decision support system expert

Alexi, Russian scientist

Nadia, Russian electrical engineer and Alexi's daughter

Domingo, drug cartel operative

Preston Abbott, National Security Council representative

Audrey "Mitch" Mitchell, Boeing technical representative

Barbara Leewon, EarthView technical representative

Frank Donovan, DEA federal finance-tracking expert; ex–New York City detective

Zipporah Moffitz, university professor, narcoterrorism finance expert, and undercover Israeli Mossad agent

Hassan Rafjani (aka "Dagger"), Iranian mullah and intelligence agent

COBI, undercover Central Intelligence Agency field agent in Iran

Paul Vandergrift, U.S. National Security Advisor and director of the National Security Council

Charlotte Adkins, former Canadian ambassador to the U.S.

Bryce Kameron, President and CEO of Intelligent Land Management Company

Admiral Stanton Lee, USN, retired; former Commander of Pacific Command, former U.S. ambassador to China and leader of the Red Team during STRATCOM wargame

Captain Dirk Baldwin, USA; Commander of U.S. Special Operations Forces (SOF) team in Afghanistan

Vice Admiral Ted Fraser, USN; STRATCOM Deputy Commander

George Tanner, Central Intelligence Agency finance specialist

Merle Beatty, satellite insurance companies association representative

General Erik "Buzz" Sawyer, USAF; Commander, Air Force Space Command

"Gunner," USA; chief noncommissioned officer (NCO) for Baldwin's SOF team in Afghanistan

Brigadier General Mike Fisher, USA; information operations expert and developer of SPECTRE 1 system

Admiral Walter Brohmer, USN; commander of both North American Aerospace Defense Command and U.S. Northern Command

Major Clint "Mannix" Gleason, USAF; B-1B bomber pilot and aircraft commander

Captain Lance "Grinch" McCartor, USAF; B-1B Offensive systems officer

Major General Donna Zurich, USAF; U.S. Northern Command director of operations

Herbert Stollack, U.S. national intelligence director

Gilbert Vega, White House chief of staff

Major Bret "Roach" Rochelle, USAF; F-15 pilot and Anti-Satellite (ASAT) mission flight leader

Major Steve "Sierra" Hilton, USAF; F-15 pilot and ASAT mission wingman

Brigadier General Lance Ferris, USAF; Air Force Flight Test Center (AFFTC) commander.

Major General (Retired) Doug Pearson, USAF; former AFFTC commander and only pilot to shoot down a satellite with an ASAT missile

1

DEADSATS

The first shots were silent.

No incendiary blast of explosives. No bombs exploding over a battlefield. No dramatic flash of a nuclear fireball to signal the onset of high-tech combat. Only a silent, single burst of electrons.

High above the Earth, drifting noiselessly in the black deep freeze of orbital space, a multimillion-dollar satellite simply died, a casualty of tiny but critical electronic circuits that failed when bombarded by a surge of electrons, exceeding the microprocessors' design tolerances. There was a brief protest of overload, a signal, then silence.

The spacecraft's final, automatic "Mayday" call—a short-burst scream that something was amiss in orbit—consisted of an innocuous stream of digital ones and zeros. Beamed to a ground station hundreds of miles below, that critical few-millisecond transmission of encrypted, coded blips would mean nothing to a casual observer. But it was the only clue that trained spacecraft engineer-detectives would receive. In its last electronic gasp, the satellite had done its part. Now its human creators would decipher the mystery, assigning meaning to brief, terminal spikes in receiver temperature measurements and power supply output voltages and currents.

Nothing new there. Engineers and technicians had done that before, hunched over computer terminals in windowless rooms scattered across the U.S. mainland.

But this time, it would be several days before those on the ground could decode the subtle, sinister messages of those last digi-words from EarthView-4 and relay their chilling conclusions: the first shots of World War III had been fired.

4 APRIL 2010/STRATEGIC COMMAND HQ/OMAHA, NEBRASKA

United States Air Force General Howard Aster frowned, yet nodded. "Continue your briefing, colonel. We'll get into the 'hows' and 'maybes' later." Interruptions from the civilians scattered around the room irritated him, and he was anxious to get back on track. He had been dreading this precise moment for months, the time when his recurring space nightmare would become hard reality.

As commander of U.S. Strategic Command, or STRATCOM, Aster occupied a high-backed leather swivel chair at the head of a long, carrier-deck–like table lined by his uniformed military staff, several gray-haired vice presidents from three commercial satellite companies, a squat and very round National Security Council representative, and J. D. Hart, a NASA technical troubleshooter. Behind them, lining the walls of a large STRATCOM headquarters conference room, sat and stood a multitude of lower ranking officers and civilians. *An odd mix of people and skills,* Aster thought as he scanned the crowd.

One of only a handful of four-star generals designated America's "Combatant Commanders," the Air Force officer wielded considerable power within the U.S. military's chain of command. Today, though, he was simply an aging former fighter pilot trying to understand jargon tossed about by an ad hoc group of staff officers, consultants, and corporate executives assembled to assess what was quickly turning into a technological nightmare. At least that's what Aster hoped it was. Because if what was happening hundreds of miles above the Earth's surface was not just a collection of random events, but something engineered by intelligent design, Aster knew he'd soon find himself in a world of shit.

At the opposite end of the table, Army Colonel Jim Androsin, a tall, thin, ramrod-straight officer, stood beside a big-screen, high-definition display that dominated an entire wall. He leaned over the long table, tapped the keys of a notebook computer, and the huge screen displayed a computer graphic of multiple satellites drifting above a crescent of blue-marble Earth.

"To recap the situation, sir," Androsin continued, "three Trans-America Satellite Company—TransAmSat, if you will—spacecraft have experienced technical problems over the past month, and a fourth had a similar anomaly early this year. TAS-5, a comsat, is the latest casualty. It appears to have a faulty battery that will force the company to shut off several transponders about one hour every day for a month during the spring and again in the fall."

"Why's that?" Aster interjected.

"Those are the solar transitions, general," responded Jack Molinero, a conspicuously well-fed TransAmSat vice president in a poorly fitting pin-striped suit and food-stained tie. His taut white shirt flowed over his belt as he slumped in a chair, trying to compose his thoughts. The tired-looking company vice-president ran TAS operations. And, today, Jack was obviously not a happy man.

"In essence," he said, "there won't be enough sunlight hitting the solar arrays each day to keep our remaining good battery charged up through the night portion of each orbit. So, we off-load the power system by selectively shutting down some of the least-critical transponders. We intentionally 'brown-out' the satellite."

Aster barely nodded his thanks and motioned with an eyebrow for Androsin to continue. Although he had been on the job as STRATCOM chief for a little over a year, the general was still getting up to speed on the finer points of this space business. As a fighter pilot, he had spent his entire career below 50,000 feet. Space, so-called Information Operations, and missile defense were new games for him—and for a lot of other people in this room, too, he realized, glancing at the faces turned toward the colonel. Back in the early 2000s, STRAT-COM had been reconstituted, absorbing the old U.S. Space Command, and was subsequently assigned a host of additional responsibilities. On top of its traditional nuclear-deterrence role, the command's bulging portfolio had been a handful for his predecessors, a fact he appreciated more each day.

Although rumors claimed that Aster had been in the running for the Air Force's vice chief of staff slot, the service's top general had asked him to take over as the nation's number-one, four-star "space warrior." Obviously, the chairman of the Joint Chiefs of Staff had already given his approval, or the USAF chief would never have offered him the job. Aster had jumped at the joint-command opportunity, preferring to remain close to front-line operations and as far from Washington, D.C., as possible. Two previous tours in the Pentagon had bred a strong dislike for things political, and it showed, despite attempts to conceal it. Because he knew it showed, he finally stopped trying to play the game. He was a warrior, not a politician.

Evidently, he hadn't irritated too many on Capitol Hill, though. After several meetings with the Joint Chiefs chairman, the secretary of defense, a slew of congressional staffers, and even the president himself, Aster's confirmation had breezed through the Senate without a hitch. The Senate Armed Services Committee Chairman's only proviso had been that Aster remain in the critical STRATCOM job a full four years. That was the same as pointedly telling him: "You'll retire in the job. This is the end of your military career." That was acceptable, though, because he was now on the cutting edge, leading the nation's most powerful combat forces. And his unofficial title—"Chief Space Warrior"—underscored the reason Aster had jumped at an opportunity to command STRATCOM. He was acutely aware that space was the new high ground of military matters, a theater of war. Now, his was a once-in-a-lifetime chance to help shape the battleground of the future, a rare opportunity for any military commander.

Unfortunately, as he stared at the table of officers and experts, listening to their somber, highly technical discussions, it was becoming increasingly obvious that the future had already arrived like a hungry wolf pounding on the front door with a vengeance. U.S. space assets were dying in orbit at an alarming rate, a pace well in excess of coincidence. And, for the moment, nobody understood why or what to do about it.

When Aster was a freshman, or "dooley," upperclassmen at the Air Force Academy had nicknamed him "Steve Canyon" thanks to his cartoon-like square jaw and blond hair. Aster was no longer that same young man. Today, the tall, prematurely white-haired STRATCOM general leaned forward, the weight of the world on his back, trying to assimilate all that the Army colonel was describing. He fought to keep his mind on the conversation in the room, but it drifted back.

Too tall to fly fighters, huh? He half smiled at the memory. *What the hell did they know?* Somebody had told him that nonsense as soon as he'd begun flight training as an Explorer Scout, while still in high school. Even the senator who'd interviewed him during his Air Force Academy application process had told him he probably couldn't fit into a sleek fighter. "Better think about bombers or transports, son," the senator had advised.

But Aster had tossed off the advice of all naysayers. Good thing he had, too. There was nothing like flight in a powerful, single-seat fighter, cruising thousands of feet above the Earth, snapping your craft into a steep bank and watching the horizon go vertical through the canopy, then pulling the stick back until g forces threatened to crush your body, trying to drive your butt through the ejection seat. *Nothing* like it. And nothing like nudging your fire-control radar's target-designation box over a hard-turning Iraqi MiG's red icon, hearing the growl of the missile-locked tone in your helmet's earphones and squeezing off an AIM-120 air-to-air radar-guided missile. Bad guys could run, but never fast enough to outrun an AMRAAM.

But that seemed a lifetime ago. Human voices intruded on the memories, pulling him back to the present. A buzz of techno-babble indicated the group was still trying to reason through whatever was killing America's eyes and ears in space.

"Two similar TransAmSat birds, Nova 4 and Nova 7, experienced failures of primary spacecraft-control microprocessors in just the last few weeks," Androsin said, pointing to a colored graphic of both satellites on the big screen. "Nova 7 is running on a backup processor, but Nova 4 lost its last backup in May. 'Four' is totally out of service now—which cost TAS the use of 48 transponders. That was most of the company's spare transmission capacity for serving the U.S. market. TAS-5, the bird with this new battery problem, serves Mexico and Latin America. Finally, TAS-6 started having problems with solar arrays last year, and that's forcing the company to gradually turn off its transponders as available power diminishes."

Androsin looked around the room and asked, "Questions about the TransAmSat birds before we move on?"

Adrian "Matt" Dillon, a no-neck, fire-plug-shaped Army colonel, the service's Colorado Springs–based space-operations commander, hunched forward, elbows on the big table. Anybody who'd followed college football in the late 1980s remembered Dillon's end-around sweep during *that* certain Army-Navy game, shedding tacklers as he rumbled like a freight train toward the end zone. The guy had never been fast, but once he built up a head of steam, legs pumping like a pair of pistons, he could drag swarms of would-be tacklers along for the ride. The Denver Broncos had drafted him, but he turned down a promising NFL career, believing his duty was to serve as an Army officer. After all, American taxpayers had shelled out for his West Point education, and he'd damn sure pay them back, with sweat-and-blood interest.

Dillon stared at Aster as if they were the only two in the

room. "Sir, to get everybody here on the same baseline, I'd recommend having Jim summarize other commercial satellite losses over the last few months, before we discuss the loss of EarthView-4."

"Good idea, Matt. Could you do that *real* quickly, Jim?" the four-star asked Androsin. The demise a week earlier of EarthView-4—a commercial, multispectral imaging satellite often used by the Pentagon to augment classified-spacecraft coverage around the world—had triggered this mass meeting, and sorting through its convoluted particulars and ramifications would take a while. Aster wanted *all* commercial satellite losses on the table before they tackled the latest imaging-sat problem.

The loss of EarthView-4 had clearly set off alarms in Washington, because the bird had constituted a critically important chunk of the nation's remaining commercial eyes-in-the-sky fleet. Its demise had hurt private-sector customers, but—and more importantly from Aster's viewpoint—it had left the intelligence community "blind" to activities in specific world hotspots. Budget shortfalls over the past decade, plus considerable political pressure to underwrite a chronically struggling commercial remote-sensing satellite industry, had left the national security community far more dependent on private-sector imaging satellites than many believed was wise.

In short, as Congress confronted a dramatic run up of oil prices prior to the first skirmishes with Iran, it had turned to military R&D and acquisition budgets, repeatedly slashing them with abandon. At the time, cutting military funds in favor of social programs had paid political dividends, but in the end, had proven extraordinarily costly in security terms. The Armed Services Committee still routinely rejected Pentagon attempts to build large, very costly but robust government-owned intelligence-gathering spacecraft, relying more and more on the commercial sector for high-resolution images of ground targets.

Damn! Now we're paying for all that stupid money-saving! Aster fumed silently. Department of Defense satellites were far better protected from all sorts of threats, designed to weather everything from radiation produced by nuclear blasts in space to sunspots. Commercial birds weren't. But they were much cheaper to build and operate than the DOD's "Battlestar Galactica" spacecraft, as the antimilitary media had dubbed them.

Aster and his staff were under considerable pressure to deliver answers—and soon—because Congress and the White House were starting to ask hard questions about what was happening. So far, he didn't have a hell of a lot to tell them, and what he did have was decidedly not good.

An Air Force major had handed a list to Androsin, who scanned it briefly, then continued. "Sir, we have EarthView's loss of their EagleEye 1 commercial sub-meter imaging satellite in December 2008. That was the company's first high-resolution image-sat, and it cratered . . . er, was lost . . . just days after a successful launch from Svobodny Cosmodrome in eastern Russia. Everything looked fine initially: stable, circular orbit and all key parameters were nominal, as were the communication links. Four days after launch, ground controllers lost contact with EagleEye 1 as a result of 'an anomalous satellite undervoltage condition,' according to EarthView. They tried to power down all noncritical equipment, then slowly recharge the battery, but failed. The bird was declared a loss a few weeks later," Androsin summarized.

"Bull!" a new voice, challenging and authoritative, bellowed.

Every head turned toward J. D. Hart, a NASA troubleshooter, who had developed a reputation for solving extremely complex on-orbit problems with government-operated civil and military satellites. Although he had little experience with commercial spacecraft, he stayed abreast of the rapidly changing technology they employed, and tracked every news account of on-orbit problems. Both *Aviation Week & Space Technology* and *Space News* reporters routinely consulted Hart about in-flight satellite glitches, because his explanations were more forthcoming, logical, and technically understandable, and generally superior to those offered by career-conscious bureaucrats. A no-spin guy whom the technical publications trusted, the casually dressed

Hart also was viewed as a loose cannon by most of official Washington. And he was showing why as irritated military brass shifted in their chairs to see who had challenged the accepted groupthink. This was precisely why Hart was definitely *not* trusted by those in power, rarely consulted by them until the last possible moment, when everybody was ducking to avoid debris splattering from the proverbial fan.

"Hell, half the people in this industry know damned well that a receiver fried itself on EagleEye," Hart growled. "And that dragged down a power supply. They ran out of battery trying to command the bird back to life. Regardless of what the company says in public, their space geeks never did figure out why the receiver nuked itself in the first place. We gotta count this one as an unknown."

Hart glared back at the somewhat stunned stares around the table, his disheveled salt-and-pepper hair sticking forward like a prickly ledge above thick eyeglasses. A pinstriped dress shirt looked as though it had skipped an ironing board on its way to and from a suitcase. Hart's knock-this-block-off-my-shoulder bearing had frightened more than a few rapacious defense contractors into revising their cost overruns before resubmitting them to the Pentagon's procurement managers. Back in the day, Hart had been the go-to guy when the secretary of defense had a budget gap to close and couldn't figure out how to do it.

Colonel Androsin stifled a grin. He had strongly recommended that Hart be included in this eclectic group, and not just for the man's broad expertise. Hart was notorious for being brash and so non–politically correct that he had been banned from testifying on Capitol Hill. NASA still needed the aging troubleshooter, but tried to keep him well away from the D.C. spotlight. Androsin liked him for the very same reason, and had found him a refreshing counterbalance to traditional thinkers during wargames. Hart called the shots as he saw them, niceties be damned, and was one of the best engineering minds in the nation when it came to sorting out spacecraft conundrums.

The NASA expert was a pioneer of what had rapidly become a valuable and rare breed: multidisciplinary engineers who could dissect downlink bit streams to meticulously determine what had triggered a satellite's death throes. A failed component? Poor systems engineering during fabrication? A burst of radiation from a solar flare, or maybe a grain-of-sand-size micrometeoroid slicing through a critical part at more than 17,000 miles per hour? Or had someone intentionally disabled the bird?

Androsin broke the awkward silence, returning to his list. "Although nobody can prove they're related, sir, there've been other unknowns, as Mr. Hart mentioned. We lost a NASA remote-sensing Landsat last fall. It went into the drink after launch, a casualty of a launch vehicle's staging failure, but we still don't know why. And the Israelis lost their Ofeq-9 all-weather imaging satellite under suspicious circumstances. But they're not talking, so we don't have much to go on. There've been several other failures and losses, like a few of the Excalibur Big-LEO comsats, but no airtight link has been pinpointed to particular causes, either."

It took a long second for Aster to recall that "Big-LEO" referred to the many satellite constellations of communications birds crisscrossing the Earth in LEO, or low-Earth orbit. Once distinctly out of favor, "Big-LEOs" had quickly become the cornerstone of a resurgent global commercial space industry in the late 2000s. Eventually, more than a hundred Excalibur spacecraft would be in orbit, bringing high-bandwidth data and anywhere, anytime voice communications to subscribers in both hemispheres. The promise of the now quaint, low-bandwidth, 1990s-era Iridium satellite constellation had finally come to pass.

Once the Excalibur space infrastructure was in place high above the Earth, customers could use tiny handheld, media-convergent personal communicators to call or access various types of data from any point on the globe. And do so without the exasperating dropped calls and fade-in-fade-out signals of first-generation cell phones, which had transformed on-the-go communications almost a decade earlier. Further, road

warriors were already able to plug their featherweight notebook computers into those handheld communicators and
have instant high-speed access to what was now being called
by geek types the "Extranet." Excalibur would expand that
capability to worldwide coverage.

Handheld communicators—combinations of personal data
assistants, palmtop computers, and digital mobile phones, all
with "nanopaper" touch-sensitive screens—had proliferated
over the past five years, combining net-cam surfing with text
messaging and conference-calling. The success of Apple's
iTunes online sales outlet in 2005 brought television and independent movies to these devices, turning WiFi-enabled
commuter trains along the Washington-to-Boston corridor
into rolling personal movie theaters. Millions of users across
the globe now routinely accessed their own private Extranets
anytime, anyplace.

Consequently, the loss of even one broadband communications satellite quickly overloaded already stressed digital
pipelines. Businesses around the globe had come to depend
on a robust, always-there satcom infrastructure, and they were
suffering now. From boardroom execs to Washington Beltway
wags and even sales reps in suburban Boise, comm-sensitive
customers were well aware that something in space was going
very wrong.

Excalibur and its copycat systems were the leading edge of
yet another communications revolution, one that could
reshape a nation's demographics, tech-journalists were saying. After all, Excalibur would enable people to live and work
wherever they chose, yet have the same high-speed wireless
access to data and electronic files they previously could only
get via their office's internal high-bandwidth local area network, or scattered wireless "hot spots" in certain cities. Excalibur and its ilk would truly be freedom-sats benefiting
millions.

IF the damned satellites are still up there and functional!
Aster grimaced, stealing a look at an oversized pilot's
chronometer on his wrist. *Two bits says the SecDef and*

somebody from the White House will call in the next few hours, and I'd damned sure better have something smart to tell 'em. He sensed that the group around this table still had a lot of ground to cover. Aster thanked Androsin for his briefing and turned to three vice presidents at his right. "Would you gentlemen care to give us a brief rundown on your troubleshooting efforts, and your educated guesses about what's going on here?" The general's tone was courteous, but his look underscored "brief."

TransAmSat's Molinero stood up and almost wearily shuffled to the front of the room, where he poked at the computer Androsin had been using. He quickly flipped through several PowerPoint slides, occasionally pointing to a graphic projected on the big screen to illustrate subtleties of often arcane, nonintuitive orbital mechanics. Finally, he faced the assemblage and carefully aligned fingers of both hands along the edge of the conference table before speaking. "The bottom line is, we can't find anything—*anything*—in the design of the satellites, the systems engineering, the prelaunch test data, or anything else that would explain the anomalies on TAS-5 or the two Novas."

"So, what's your best guess, Jack?" Aster urged, after a strained-silence pause.

Molinero didn't relish what he had to report. Aster noted how the man shuffled his weight from one foot to the other before continuing, appearing to consider how the group would accept his conclusions. "General, we think . . . ," he hesitated, lips pursed, "we think somebody is interfering with our birds. No other explanation makes sense."

The room's skepticism was palpable, hanging in the air like steam.

"What evidence supports *that?*" The question came from a one-star general, the U.S. Air Force Space Command (AF-SPC) director of operations, who had flown to Omaha that morning for the troubleshooting session. As the command's senior representative, he spoke for the bulk of America's military space force. The Army and Navy also boasted space commands, but the USAF's was the largest. The service had been tapped as the nation's executive agent for military space based on an investigative commission recommendation. And that one-star was now responsible for "Space Control," a mission that encompassed protection of spacecraft, military or commercial. *This crap is happening on the good brigadier's watch, and he isn't happy*, Aster smiled watching the one-star squirm.

"Only circumstantial. No proof," Molinero conceded. "On the other hand, to be frank with you, sir, we're worried sick that we'll lose more of our birds. And you Air Force guys had better do the same! Yours could be next!" he barked at the general.

Aster smiled, unable to resist an opportunity to lighten the darkening mood. "Jack, this isn't an inquisition. Take it easy!"

A titter of uncomfortable laughter fluttered around the table, and Molinero winced. "Yeah, yeah. I guess we're all in this mess together. But we'd better figure out what the hell is going on, and try to stop it before we collectively go completely deaf and blind in space. Look, when Nova 4 died a few years ago, something like 43 million pagers in the U.S. went stone cold dead. Doctors couldn't get emergency pages inside their own hospitals, for crissake! The loss of just that one satellite, Nova 4, knocked out banking networks, automatic teller machines, gas-pump credit-card verifications, and a few specialized corporate comm networks."

Molinero paused to let that sink in. "As you all know," he continued, his hand sweeping the room. "Every one of the spacecraft this industry lost around that timeframe had the same kind of symptoms that have been discussed here today, at least what we've talked about so far. You might think we're just revisiting an old situation. But this time, my gut tells me it's not a manufacturing—"

Aster interrupted. "You're right, Jack. This smells different. I think we all agree on that," he said, standing abruptly.

"Look, I have to step out to make a call, so let's take a ten-minute break. My handheld *is* still working, and it's squawking right now." To polite laughter, the general nodded and headed for the door, a muscular uniformed aide in close trail.

Aster's communicator text message flashed the number of a staffer for Senator C. I. Creighton, an outspoken populist from the decidedly liberal Santa Monica area of southern California. She was demanding to talk to the STRATCOM commander right *now*.

This had better be damned important, Aster thought, already more than irritated by the interruption. Politicians were not his favorite people, and Creighton's elitism and arrogance were particularly distasteful. Besides, Aster had a pretty good idea what the pompous chairman of the Senate's primary military space or "milspace"oversight committee wanted to know.

Minutes later, a land-line phone receiver clamped to his ear, the general's jaw muscles were flexing in double time. Always an unmistakable indicator that an Aster mini-explosion was in the making, his aide noted. The blue-suited lieutenant colonel stood and silently motioned that he'd leave the spacious, yet simply furnished STRATCOM commander's office, ensuring the general's privacy. Aster impatiently waved the aide back to a leather couch. Lt. Col. Thad "Burner" Burns, a powerfully built African American B-2 Spirit bomber pilot turned STRATCOM commander military aide, sat and pointedly focused on sorting a stack of paperwork he always carried, trying to feign he wasn't hearing the general diplomatically skin some young, self-important Capitol Hill staffer.

Burns concentrated on the memos that Annie, Aster's executive assistant, had given him for the general to sign before the day ended. This was always a challenge, given the boss's grueling, nonstop schedule. The ex–bomber pilot hated this memo-carrying part of the job, but it went with an aide's territory.

Thad Burns's "real" job rested next to his spit-shined, low-quarter right shoe. In that innocuous briefcase, Burns carried the no-joke, go-to-war codes that Aster would use to launch the nuclear holocaust that every soldier, sailor, airman, and marine in uniform hoped and prayed would never happen. If the president's military "shadow," a Burns counterpart who never strayed far from the commander in chief's side, ever opened the "football" he always carried, and the president used its contents to transmit a nightmare "go" message, Aster would follow suit. Consequently, Burns and his innocuous "football" were Aster's shadow, and that proximity meant he carried directives, memos, and other bureaucratic make-work paper to be signed, as well.

Tagging along behind a four-star wasn't exactly what Burns had expected when he signed up for flight training years ago. Given a choice, he'd rather be up there, at the edge of space, winging his way to a target, his B-2 Spirit deflecting enemy radar pulses and remaining a phantom, invisible to electronic probes. But the Air Force's arcane assignment system had seen fit to put this "football" at his feet. Friends said that it meant he was one of the five most trustworthy people in today's military. Maybe, but he wasn't convinced.

"Miss Peloni, let me first clear up a critical point, then I'll try to answer the senator's specific questions," Aster stressed, his tone icier now. "The United States—in fact, the whole developed world—is incredibly dependent on space these days. There's something like eight hundred billion dollars' worth of hardware in orbit now, and more's going up every week. We have a huge financial investment in space, and that investment has now become vital to our national and economic interests. Space *is* a national center of gravity, contrary to your last comment. And I'm sure the senator appreciates how absolutely critical our military space platforms are. Without them, this nation simply cannot fight a major war today. They're critical force multipliers, compensating for the huge cuts our forces took in the nineties, after the Cold War ended. Space is our eyes and ears, our link between command centers here in the States and forward-deployed forces throughout the world. It's the heart of our missile defense capabilities, too. It's absolutely essential to our ability to navigate—to find our way through jungles and deserts and across oceans—and to guide precision weapons without causing . . ."

Interrupted, the general nodded a couple of times, but those jaw muscles twitched even faster. Four-stars weren't accustomed to being interrupted, and really didn't appreciate it when done so by young, self-important Washington lightweights, Burns knew. The aide now held his breath, waiting for the explosion he expected to come, oftentimes aimed at his own head.

He'd heard Aster say the same things about space during recent speeches, expressing a surprising level of candor for someone in his position. Before he'd taken over as STRATCOM chief, few uniformed space professionals in recent years had dared utter the politically volatile words "space control" or "space superiority" in public forums, fearing a predictable backlash and outcry from powerful congressional and European doves who clung to the amusing, naive view that Earth-orbital space could forever remain free of conflict.

The current White House had displayed surprising pragmatism as well as guts, Burns thought, by making it very clear that the ultrasensitive space superiority mission had been danced around long enough. Aster, as the top U.S. space commander, was responsible for it and other, more traditional space-support activities, and it was time to reduce theory to practice, the president had said. Until that directive, space superiority had been largely ignored by Washington—and STRATCOM, as well, because there had been no political support for working the issue. Although Air Force officers had been sounding the alarm for years, and Congress knew it would be important someday, "space superiority" had simply been relegated to the too-hard-to-do category, much as the threat of terrorism had been kicked downstream prior to that horrible day in September 2001. However, that lapse in space-related concern was starting to bite the nation in its overexposed butt, Burns thought.

"Yes, ma'am. You can tell the senator that, one, we're narrowing the range of possible causes for EarthView's loss, and I expect to have something for him and the president within a day or so. Two, we have an excellent team here, working very hard, going through a logical process of assessing several potential causes. . . . I'm sorry, say again? I didn't catch that . . ."

The general glanced toward his aide, waving him over. Quickly scratching a note, Aster ripped a light-blue sheet topped by four stars from a pad and shoved it across the desk. A.—GET SPEED @ WH HS ON PHN!! it read. Burns stifled a grin and headed for the door. Creighton's staffer had pegged the boss's "delta-sierra" meter, pilot talk for "dumb-shit" political nonsense exceeding the general's daily limit. Annie, the general's perennially poised, ultra-efficient gatekeeper and executive assistant, took no time in rescuing her STRATCOM-commander boss. Burns returned and pointed to the phone's base unit, where a new light blinked.

"Miss Peloni, I'm terribly sorry, but I'll have to . . . Crap, the White House?" Aster feigned surprise, winking at Burns. "Okay; I hear you. Please tell Senator Creighton that I'll definitely get back to him as soon as we have something nailed down here. . . . Absolutely. Always great to hear from the senator . . . Roger that. Will do, ma'am. Bye." The general punched another button, switching from the senator's staffer to a National Security Council staffer on hold. This one, in stark contrast to Creighton's Ms. Peloni, Aster knew and trusted implicitly. After all, the two generals had flown F-16 fighters together when they were mere fast-burner captains with the "Wolf Pack" wing at Kunsan Air Base, South Korea.

Brigadier General Hank "Speed" Griffin hadn't climbed into star ranks quite as quickly as Aster, his former flight lead, thanks to a few "lost" years at the USAF's Test Pilot School and as a test pilot assigned to a "nonexistent" Air Force facility in northern Nevada. The still highly classified "system" Griffin helped develop there had only recently been revealed to Aster, a system the latter just might need, and damned soon, the STRATCOM commander reflected. The president's National Security Advisor had spotted Speed's talents and tapped the one-star for NSC duties a few months ago.

"Speed! Hey, thanks for rescuing me from that little twit. . . . Say again? . . . Oh, one of Creighton's dumbass staffers. She and her boss didn't give a flyin' frap about space until EarthView-4 went quiet, and now they're demanding action ASAP. Listen, how about making sure the boss and your NSC folks stay abreast of what we're doing here, 'cause a few Senators on the Hill are apparently going to use these space glitches as political ammo against the administration. I don't know what they're up to, and I don't give a damn about stupid political games—you know that—but I can't let the commander in chief be blindsided with a shit-storm from Congress. If he is, that'll really slow us up out here, and I damned sure can't afford to deal with that B.S. right now. We may be—it's not certain yet—but we may be in some serious space doo-doo, bud. I need you to buy us time to sort this shit out."

Aster was pacing across his carpeted office in long, slow strides, one hand jammed deep in a pocket of dark-blue slacks. He quickly gave Griffin a capsulized update of what the mixed group of experts—probably now returning to the conference room down the hall—were discussing, and what Aster hoped to do in the next hour or so.

"Just buy us some time to get a plan together out here, okay? . . . Thanks, Speed."

He clicked off, slipped the phone into its cradle and headed for the door, briefly checking his small handheld communicator. Aster grimaced, but jammed the device back into its belt-mounted holder. Whatever the general had seen on that colored mini-screen, Burns noted, it was going to wait. For a few months, the boss had refused to carry one of the new do-everything communicators, a quirk that aggravated *his* bosses in D.C. The Joint Chiefs chairman had finally "suggested" the STRATCOM general "get with the twenty-first century" and start carrying a communicator, like all the other combatant

commanders did. Aster complied—hell, he really had no
choice—but chose to ignore the damned thing whenever he
could.

"So, where do we go from here?" Aster asked nobody in particular a few minutes later, while settling into the high-
backed swivel chair in his conference room. All the key
players had returned to the long table minutes earlier.

"Maybe we should tap some of the national sources, just to
see if the spooks have anything to support Jack's theories," the
STRATCOM operations director, Army Lieutenant General
Dave Forester, suggested. "National sources" was still an accepted euphemism for "spy satellites," electronic snooper
aircraft and other information-gathering platforms operated
by super-secret U.S. intelligence agencies, or "spooks," a term
intel types actually relished as a strange badge of honor. Aster
agreed, gave a clipped order to an officer standing nearby,
then turned back to the group.

"Folks, I'm being asked some hard questions. I think we'd
better start narrowing the potential causes of, and options for
dealing with, whatever's going on out there in orbit," Aster
said. "Because of the short timeline we have for providing
answers to people back east—and what, to me, sounds like
a huge matrix of 'maybes' and 'could-bes' we're dealing
with—I want to start homing in on specific issues, but doing
it in parallel. We have to take the fastest route to get at the
heart of these problems, so let's attack 'em all at once. Let's
get some multidisciplinary teams going here, spreading the
expertise you all bring, then focus on developing causes and
options for response. If you think we can rule out basic technical problems, then zero in on Jack's theory that someone is
intentionally taking these birds down." He studied the big
chronometer briefly.

"Let's reconvene at 0730 tomorrow, but in the STRAT-
COM Wargaming Center. And we need to bring some
answers, okay?" Aster's hard blue-eyed stare swept the room,

taking note of more than a few bobbing Adam's apples. The general stood, prompting a flurry of olive-drab, white, khaki and blue uniforms to jump to attention.

"Carry on," he clipped, brushing past Burns, who held the room's door open.

5 APRIL/STRATCOM WARGAMING CENTER

Colonel Jim Androsin surveyed the few people scattered around the tiered wargaming amphitheater, a cross between a college lecture hall—but with computer consoles at each station—and a high-tech multimedia facility. Androsin was mentally checking off what he knew was already there, but kept scanning the hall just to make sure. He sipped a steaming cup of Don Francisco's Vanilla Nut coffee, hoping it would clear the dull fuzz of fatigue that had settled between his ears. Aster's crash effort to explore high-potential causes of seemingly unrelated satellite losses, as well as options for replacing the military and commercial communications bandwidth, imagery, and classified capabilities the nation had lost, had forced Androsin's STRATCOM wargaming division to scramble. Unlikely as it might seem, his people were the experts Aster had tapped to work up feasible scenarios—hypotheses, really—for the growing number of satellite losses, because that's what they normally did in creating major command-level wargames. Last night, though, their talents had been challenged to the corners of their abilities to frame evolving real-world problems into reasonable scenarios. Working closely with experts from the companies, NASA, and the intelligence community helped, but the knowledge that this was no longer a "game"—it had become the real thing—added considerable stress to the entire process. This time, there were real-world consequences.

Working through the night, aided by J. D. Hart and the commercial companies' technical experts, Androsin's wargaming specialists had formulated four logical scenarios that might explain why a number of satellites had gone dead.

Fortunately, the division had already been planning a wargame with astonishingly similar precepts for weeks, a response to a growing command-wide concern about how, in the future, they'd handle precisely what was occurring now: a surge of satellite fatalities accompanied by little hard data for troubleshooting. Consequently, they had drawn some key ideas from the "Deadsat" play book, a wargame that had been conducted years earlier. But the margin for error of a traditional wargame had vanished. America's space resources, they'd learned, could be in serious jeopardy.

After a brief discussion, the mixed group of bleary-eyed commercial, civil, and military space specialists had broken up into small knots of people now scattered around the center. Arriving promptly at 0730, Aster had quickly set the stage for the day's techno-sleuthing, reemphasizing the pressing need to get at the heart of admittedly knotty problems, then turned the floor over to Androsin for a status update. Nothing had improved overnight, the colonel had reported. If anything, there were even more unknowns to deal with, thanks to commercial companies' engineers diligently scrutinizing those last-second data streams transmitted by dying spacecraft hundreds of miles above the Earth. Now the brain-numbing, grind-it-out, top-level problem-solving was under way.

Aster joined one group huddled around a powerful Silicon Graphics high-definition workstation. "I'd appreciate getting a better picture of the geometry, timing, and other factors that might be involved in these satellite losses," the general said. A young civilian contractor nodded, then maneuvered his electronic stylus, tapping a screen to open successive colored windows on the display. A bar of data and virtual control buttons lined the screen's left side, bordering an image that appeared to put the viewer high in space, looking back toward a slowly rotating Earth.

"This is an Analytical Graphics STK visualization of where satellites were when most of the failures were detected," the young contractor explained, shifting the viewing angle effortlessly. "These white ovals on the Earth's surface, like, show geographical areas that had those satellites in sight when we . . . ya know, when the space-birds died. You can see they cover millions of square miles, above and below the equator . . ."

"What's the green patch right there?" a Navy commander interrupted, pointing to the screen's center.

"That's where all the white satellite-in-sight patches intersect," the contractor said. "Now, if we show what areas of the sky were visible from that green area, and do some time correlation . . . ," he paused while a few more clicks adjusted the image, "we can get a feel for, ya know, where any interference might have come from—if there *was* any interference, I mean," he added, recalling Androsin's admonishment to stay open to any possibilities. The patch of green now covered an area of the Caucasus, reaching into northwestern Afghanistan.

"Can you blow that green part up?" asked the Navy commander, leaning closer. The contractor scribed a circle around the area, tapped the screen, and several observers nodded slowly as the commander voiced what they all noticed. "Covers some of Afghanistan, Tajikistan, and . . . whatever those other 'Stans are."

Aster's communicator vibrated again. A glance at its glowing screen caused the tall general to stiffen and those jaw muscles to tighten, Lt. Col. Burns noticed. Aster turned, made eye contact with his aide and stood. Burns had seen that look before. *Oh shit. This can't be good,* he thought. The "football" suddenly seemed heavier as he followed Aster from the room, vaguely aware that the general hadn't bothered to explain his abrupt departure. Members of the group he'd been with seconds earlier exchanged puzzled glances.

After a short break, the small groups reconvened in the center and again turned their attention to Androsin, standing at a podium this time. Aster had tapped him to facilitate the session in the commander's absence. The colonel's face was grim, lips pressed to a thin line. "Okay, people, here's the situation. General Aster was just informed that, over a six-hour period last night, we lost several military satellites. I can't share the details here, but suffice to say that some very, *very* important capabilities have been lost. Let me recap the situation now, as a result of this latest information." A list of bulleted items appeared on the big screen; he circled the first with a transient ring of red from a laser pointer.

"Since yesterday, two more commercial satellites have died. One is an Excalibur bird, the sixth lost since the constellation was completed last spring. The company was still checking out the system in preparation for starting commercial service in the fall. No big deal from a business sense, they say. Excalibur already has spares in orbit and a few more were being readied for launch. But the fatality rate is higher—*much* higher—than expected at this stage, and the circumstances around the latest failure are puzzling, at best. You'll get more details from the Pantera Corporation rep after I'm done here," the colonel said, nodding to Rob Joaquin.

"The other one is, I'm afraid, more bad news for Boeing: TVBS-1. Boeing built it, and it's operated by SkylinkTV, another Boeing-owned outfit. That bird lost a primary satellite-control processor early this year, and is still in business, operating on a backup processor. But the cause of that processor going tango-uniform is still being investigated."

"Pardon me. Tango-uniform?" asked a diminutive woman in the center's second row, her hand in the air to catch the briefer's attention.

Damn! That was dumb, Androsin chided himself. "Uh . . . well, that's military-speak for . . . uh, a generic system failure, you could say." The colonel colored slightly, acutely aware of controlled smirks scattered around the room. "Tango-uniform" was the airman's phonetic spelling for the letters "T" and "U," but, over the decades, had become synonymous with "tits up," shorthand for something going seriously wrong with an airplane. As aviators moved into key space-operator positions, the politically incorrect term had followed, despite attempts to squelch it, especially when in the company of civilians.

"What's the impact of all these losses?" a STRATCOM-assigned Marine Corps colonel asked, pointedly changing the subject.

Androsin noted the lady appeared satisfied with his vague answer, then focused on summarizing obvious ramifications of the latest spacecraft losses. What he didn't relay was Aster's gloomy assessment. Only a few minutes ago, the STRATCOM chief had verbalized what Androsin and other senior officers had suspected for several days: The United States' space assets were under attack, and the national security community didn't know why, or who was behind those attacks. But, without question, U.S. forces in certain areas of the world were already blind and deaf, unable to know—for sure—what potential adversaries were up to. The loss of intel, military, and commercial space assets was having a serious, adverse impact on the nation's warfighting capabilities.

General Aster's space nightmare was coming to pass.

2

ZEROING IN

Staff Sergeant Sean Cantrell adjusted his night-vision goggles or NVGs, bringing the yellowish scene and several scurrying figures into sharper focus. *No question; the bastards are loading those new, more powerful, and deadly Chinese-built rocket-propelled grenades into that old, battered truck.* He turned and tilted back his head to peer under the NVGs suspended a few inches in front of his face. The three special operations troops accompanying him were still in position.

"Bogey," barely visible in the ink-black darkness a few meters away, gave Cantrell a thumbs-up. He'd seen the RPGs, as well. Beyond, another camouflaged figure melted into the desert sand, lying prone and scanning the team's left flank. The young sergeant knew his team's fourth guy lay on the right flank, also watching for intruders that could instantly ruin their evening's outing.

Officially, the U.S. Special Operations Forces team wasn't here tonight. This was a deep-black covert mission inside a sovereign nation, so contact with the natives had to be avoided. And getting their tails captured was not an option. War had not been declared on Algeria, but its naive leaders had turned a blind eye to several terrorist cells that had fled Iraq several years earlier and moved into remote desert regions of northern Africa. A special U.S. envoy had quietly, but not so subtly, warned Algerian leaders about the cells, but her proof—high-quality satellite images—had been dismissed as U.S. attempts to insult Muslims yet again. Four years after cartoons of Muhammad surfaced in Danish newspapers, incensing Muslims across the globe, the Islamic world still seethed with anger. As a result, Algeria ignored the infiltration of terrorists, allowing cells to replicate like a virus, breeding yet another generation of jihadists.

Something had to be done about them, and soon. So, American and British leaders had jointly decided they could no longer afford to stand on political ceremony. They had quietly released off-the-books commando teams to roam nations that gave refuge to those terrorists, taking out cells wherever they were found.

Tonight's "black ops" mission was the new U.S. president's way of sending a non-public message he knew would be received throughout northern Africa: If *you* won't root out these rats, we will. U.S. spy satellites had spotted what analysts believed was a terrorist supply center, and President Boyer had green-lighted a Special Operations Forces mission to "take it out"—but only after positive identification. That's why Cantrell's team was here, on site: to ensure no embarrassing mistakes were made. Before conceding to this mission, the president had made it clear to the Pentagon: "There will *not* be any dead Algerian kids on al Jazeera TV. Got it?"

Cantrell aimed his AN-46, the newest handheld target-designator/communicator that the Army's innovative engineers back at Fort Belvoir's night-vision lab had developed, ensuring its small screen captured what his NVGs had revealed a moment earlier. The pint-sized designator/transmitter system first painted its target with an invisible infrared floodlight-sized beam, then did its signal-processing magic to turn dark to day. *Sooooo cool!* Cantrell smiled.

Several bearded young men wearing dark T-shirts and loose-fitting pants ran to and from a crumbling mud hut, each carrying two of the distinctive pointed, bulbous warheads, one on each shoulder. The canvas-covered truck bed wasn't visible, but movements by two men stacking the weapons verified Cantrell's first impression. A hell of a lot of the damned things were already in there.

The SOF team leader steadied his blunt-nosed, gun-like AN-46 on a hummock of hard-packed sand, fixing razor-thin crosshairs over the truck bed. He squeezed the trigger one notch, firing an invisible laser that measured the distance between Cantrell and that truck. Multiple digits—GPS satellite-derived coordinates of the truck's position—instantly appeared on the small screen. He then squeezed again, depressing the trigger further. At the second detent, those digits flashed momentarily, then steadied. Cantrell released the trigger, but kept the crosshairs in place until a small box in the screen's upper-left corner appeared, then filled with four letters: RCVD.

"Okay. Bird's got it," he whispered into a wire-thin boom microphone at the corner of his mouth.

"Show time!" Bogey breathed in response, barely audible in Cantrell's earpiece.

ALTITUDE: 17,500 FT. OVER AL QAHR, ALGERIA

USAF Captain "Pepper" Malloy noted the green-glowing GPS coordinates appearing on his head-up display, linked by a stubby line to a tiny, four-sided box. The SOF team had found it! They had a bona fide target. Malloy, assuming his wingman, "Shark" Fisher, had received the same transmission, stood his F-22 "Raptor" on its left wing and pulled, grunting as the 5-g pressure tried to crush his body. Their target was at the stealth fighters' seven-o'clock position, a good eleven nautical miles behind them. They'd be well positioned in a matter of seconds, Malloy calculated, shoving his throttles up to military power. He monitored the airspeed rapidly building, then tugged the twin throttles back, ensuring the Raptor stayed subsonic. No need to warn the bad guys with a Raptor-created sonic boom. The F-22 had been selected for this mission because it was practically invisible to Algerian radars, and had the combination of range and speed needed to get in, hit its target, and get the hell out of town without anybody on the ground ever knowing it had been there. Tonight, another band of terrorists would meet Allah face to face, aided by the business end of satellite-guided smart bombs.

Malloy thumbed the Master Arm switch and verified that the fighter's stores management system was talking to the 250-pound Small-Enhanced Joint Direct Attack Munitions mounted on trapeze-like racks inside the F-22's still-closed bomb bay. A slight grin spread beneath his oxygen mask. Here he was, flying a no-shit covert combat mission, about to drop extremely high-explosive death and destruction on a bunch of bad guys who were committed to killing American, European, and Israeli citizens, but his mind had flashed to a comment a former Air Force chief of staff had made during a first-delivery ceremony. Referring to the new super-accurate, small-diameter E-JDAM bomb—which featured both laser and GPS-guidance kits and awesome blast power—the chief had quipped: "E-JDAM—that means '*Extra*-J-DAMNED good!'"

Tonight's detailed mission plan called for Malloy to drop two E-JDAMs during the two-aircraft flight's first pass over the target. Shark's jet carried a couple of backup E-JDAMs, just in case. With dozens of training-weapon deliveries under his belt, Malloy had no doubts his two blivets would evaporate whatever the SOF team had discovered down there. The bombs' GPS-guided accuracy and killing power were absolutely uncanny. Thanks to a covert ultrawideband signal the SOF troops had beamed to a ghostly comm-relay satellite cruising above Algeria's desert, three-decimal-place GPS coordinates had been loaded automatically into the bombs' guidance system. Malloy's job was simple: fly his jet to the right spot in the sky, then let the cosmic Raptor computers do the rest.

Careful refinement of an image obtained by an ultra-brief blast of energy from the F-22's advanced electronically scanned array radar in the fighter's nose showed a cluster of buildings thousands of feet below. But a target-designator box on the heads up display was nowhere near the buildings. Inexplicably, it rested over what looked like nearby lumps of desert. Malloy bumped a switch on his control stick, zooming the stored radar image closer, and saw the TD box hovering near several faint blobs on the ground. Something didn't feel right.

Tapping a keypad on the forward instrument panel, Malloy cross-checked the new UWB frequency, making sure it matched one printed on the mission card strapped to his right leg. He hesitated, not really wanting to break radio silence. But he needed to know for sure. *This ultrawideband covert-radio shit had better work as advertised.* He grimaced.

"Cotter 23, Musket One here. How copy?"

"Musket One, Cotter 23. Copy five-by. You in the area?" Cantrell whispered, surprised to hear the pilot's voice. His team had been briefed that this would be a no-comm mission all the way. The magic of datalinks, space comm-relay platforms, and microsecond bursts of covert UWB signals were supposed to preclude any need for old-fashioned voice communications.

"Rog. Confirm your position and that of the target, Cotter. I'm getting a major discrepancy here." Malloy rechecked his navigation system setup for the fifth time. The configuration looked good, but his inertial nav position didn't agree with the displayed GPS position. The two usually matched dead-nuts-on, but not now. The inertial system placed a HUD-displayed target-designator box close to a vehicle parked near one of the dilapidated huts. The pilot could see what he assumed were people milling around that vehicle. But when he switched to the GPS-only TD box, it hovered a hundred meters to the west, near a handful of spots detected by the radar system. Those spots weren't moving. Overall, not good. The inertial and GPS-generated positions should coincide. Of course, inertial systems could drift over time, but he'd rarely seen that happen in the F-22. As long as the numbers were close to being the same, the inertial always defaulted to the GPS-designated position. Now, they were well-separated for some reason. That meant they disagreed, and by a significant amount. But which was correct?

On the ground, Cantrell swore under his breath. *Sonofabitch! Terrorist bastards in our sights, and some anal Air Force wuss is gonna screw around and let 'em get away— and with a boatload of friggin' RPGs!* He punched his advanced Garmin 8F nav system, cross-checked the display against that of the AN-46 and whispered a three-decimal-place set of GPS coordinates into his microphone. He then fired the AN-46's laser/transmitter at the RPG-laden truck, beaming yet another set of target coordinates to the fighters.

Overhead, Malloy carefully wrote the strings of figures on his kneepad, then compared them with those on his HUD. The SOF team's GPS coordinates were several hundred meters west of the GPS-positioned target-designator box, which still hovered near those stationary spots on the desert ridgeline—right where the SOF team said the target was. Malloy concluded the team was nowhere near the target, safely out of E-JDAM range. *Okay! I'm not sure you're on that ridge, terrorist dudes, but you'd better get ready to greet Allah and those virgins.*

"Rog, Cotter. I show you well clear of the target. We have the bad guys in our sights, and we're going to light 'em up," Malloy radioed.

"Cleared in, Musket! And let's get on with it, okay? They're about to skip—and we need to bug out, too," Cantrell whispered. Daylight wasn't that far off, and his team had a long hump to its extraction point. The MV-22 Ospreys were already on their way, and he did *not* want to miss a rendezvous and ride home with those tiltrotor aircraft.

Just to satisfy a gut-level uneasiness that persisted, Malloy tapped a quick datalink message and hit ENTER. Seconds later, Shark's two-word response appeared on the head-down display: CNFM-SAME. Satisfied that the GPS-designated target he and Shark had identified independently was, indeed, the bad guys' location, Malloy maneuvered the jet's nose slightly, adjusted altitude and pressed the "pickle" button on his F-22's stick. Computers took over, making thousands of calculations per second until the fighter's electronic brain said "Now!"

The pilot heard and felt his Raptor's weapons-bay doors pop open and two support pylons extend into the airstream. Twin bumps and a slight twitch of the flight controls signaled that both E-JDAMs were off and headed for their target thousands of feet below. The bay doors snapped shut, confirmed by indicators in the cockpit, and Malloy exhaled. The Raptor was back in stealth mode again. For the brief time those bay doors were open, his fighter could be seen on radar, but no longer. Again, the F-22 became part of the night sky, protected by the best antiradar stealth system in today's world.

Turning about 30 degrees, Malloy monitored time-to-go numbers ratcheting down on his HUD, mentally tracking the two bombs as they hurtled Earthward, guided by signals from Navstar GPS satellites thousands of miles above. A bright double-flash in the night, far below, told Malloy his bombs had found their target. The GPS-driven target-designator box on his transparent HUD lay smack over that bright spot. Shack! Bombs right on target!

On the ground, five heads snapped to the west, eyes wide. Huge explosions a hundred meters away, near a low ridge, stunned the young men loading RPGs, triggering a flurry of shouts. In Arabic, a leader screamed orders, prompting all five to leap into the truck.

Seconds later, as the blossom of twin detonations faded somewhat, Malloy's highly detailed radar display showed tiny digitally processed blobs racing to that vehicle parked near the shacks. The pilot thumbed his radar's cursor until it overlaid the vehicle; zoomed-in, then bumped the lock-on switch, ensuring the radar would auto-track the truck, regardless of the jet's movement. He saw a ghostly image of an aging vehicle jerk forward, then accelerate, swinging onto what passed for a road through the sand.

"Cotter, Musket One. What the hell is going on down there? You tracking that truck?" Malloy demanded via radio. Nothing. No response. The pilot's gut tightened. His eyes jumped from the truck to that bright, searing fire still under the hovering GPS TD box on his HUD. The truck raced away at an increasing rate, bouncing crazily.

"Cotter 23, Musket One! Do you copy?!" Malloy repeated, sweat starting to run down his forehead. Again, nothing.

"Oh shit!" Shark's hushed voice appeared in Malloy's helmet. "Did we . . . ?" Malloy's heart seemed to stop beating. How the hell could . . . ? No! No way! The damned GPS never lied! Inertial nav systems would drift with time. You couldn't trust those bitches. But GPS . . . it was *always* dead-on!

Shark recovered first, voice tense. "Pepper, I think the bad guys are getting away. Suggest we do something, fast."

Malloy, brain spinning, shook his head, trying to clear the shock that threatened to consume him. *Oh, God! Please, let's go back and do this over . . .* This couldn't be happening!

"Musket One; you with me?" Shark's sharp tone seemed to slap Malloy.

"Uh, yeah. With you. Uh, let's . . . you've got the lead. I'm winchester," Malloy mumbled, indicating his bomb load had been exhausted. A double click in his earpieces acknowledged. The faint outline of Shark's F-22 eased forward until its left wing appeared. Malloy automatically adjusted his jet's position, now flying as Shark's wingman. A shallow turn and Shark was wings-level again.

"Musket Two's got a lock. Refreshing the radar . . . Readyyy . . . pickle." Malloy saw Shark's weapons-bay doors snap open, two mean-looking, dark objects extend, then fall away. Doors banged shut and Shark banked gently to the right. Malloy glanced inside his own cockpit, satisfied that his radar was still auto-tracking the truck. Time seemed to stop. Only the Raptor's engine-rumble filled the night for what seemed like minutes. Then a massive explosion drew both pilots' attention outside. A huge fireball, punctuated by bright arcs spraying in all directions, lit the sky, reminding Malloy of a fireworks display. Secondary explosions rippled across the desert as RPG warheads scattered, flung clear of what had been the truck's image on Malloy's multifunction display.

Even from this altitude, the truck's fiery disintegration was awe-inspiring.

"Nice hit, Shark. I've got the lead," Malloy said. Shark noted the dark tone in his flight leader's soft tone. This was going to be a very long trip home.

Still unbelieving, Malloy tried one more time. "Cotter, Musket One. How do you read?" Again, only silence. Drawing a ragged breath, Malloy switched to his alternate radio and made the toughest call of his life. The two inbound MV-22 tiltrotor Ospreys would be diverted to a new pickup point, one near a cluster of now-abandoned mud huts in the Algerian outback. He doubted whether they could get there before dawn, significantly increasing risks for those Osprey flight crews. He also feared they wouldn't find enough to warrant landing. But he knew they'd land anyway. The SOF never, *ever* left their dead behind.

6 APRIL/2ND SPACE OPERATIONS SQUADRON, SCHRIEVER AIR FORCE BASE, COLORADO

Staff Sergeant Mitch Kucera's forehead furrowed as he tried, for the third time, to transmit a burst of digital codes to a Navstar GPS satellite. He'd hoped to send the update package in a routine burst transmission within seconds of the navigation spacecraft peeking over the horizon, finally in view of a big-dish antenna that shadowed a ground-based control station in southeastern Australia.

Normally, those few milliseconds of digital data would have been snagged by the five-year-old Block IIR-M1 satellite as it coasted silently in mid-Earth orbit or MEO, 12,500 miles above the planet. One of twenty-four active spacecraft that comprised the GPS constellation in 2010, this one was due for a tweaking of several onboard parameters, ensuring that positioning and timing signals it continuously beamed to the Earth were accurate to a gnat's eyebrow. Kucera's job was straightforward: assemble the needed error-corrections on one of a dozen computers scattered throughout the 2nd Space Operations Squadron (2SOPS) command center, then click on a TRANSMIT icon. For the few years Kucera had worked in 2SOPS, that simple click of a mouse had been followed seconds later by a familiar acknowledgment that the bird in orbit had received its stream of digital bits, automatically updating critical navigation parameters. But not today. Kucera waited, anxiety inching up a notch, but the screen's cursor continued to flash, frozen in the STANDBY position.

Grimacing, Kucera reached for a phone, eyes still locked on the unmoving, flashing STANDBY. As he waited, the four-stripe sergeant took note of an uncomfortable feeling. His gut was predicting a very busy, not-so-good shift today.

"Sir, Sergeant Kucera. We've got a problem that I think you should look at. One of our birds won't take an update. . . . Yessir. I sent it three times. No acknowledgments from the 'sat' each time. Nothing at all. . . . Yessir; ephemeris is fine and all payload health parameters look good on the downlink. There's just no indication the update's been received. Until today, I've never seen anything like this. And . . . uh . . . sir, according to the log, this is the fourth bird in the last twenty-four hours that's rejected updates. . . . Yessir. Same thing happened three times on the night shift. . . . No, sir. Four different spacecraft."

Minutes later, Kucera and his Second Lieutenant shift leader were convinced something was seriously wrong. Checks of other in-view Navstars showed that four responded to uplink transmissions, acknowledging as they were supposed to. Another two now in view of the Australian ground station would not. Overall, there were four incommunicado GPS birds—at least in the receive mode—up there. All four continued to transmit navigation and timing data, though.

"Better get the contractor guys in here," the young officer said. "I'll notify folks upstairs. Give me a yell on my hand-held if you finally get through to any of these birds."

"Roger, sir," Kucera said, noting the communicator's number was already written in his logbook. He then ran an index finger down a plastic-covered list of five-digit numbers. "We're out of the window for that first one, but two of our no-response birds will be in for eight to ten more minutes. Then we're out of luck for another couple of hours." The lieutenant was relatively new to the squadron, Kucera recalled. Wouldn't hurt to reinforce techie details that the shift supe might need when the big bosses upstairs started asking hard questions.

Hours later, Kucera had more senior officers huddled around his workstation than he really wanted or needed, some hovering over his shoulder, eyes locked on the computer screen. Others talked quietly among themselves. However, not a damned one of them had any answers that made sense, Kucera noted. He glanced at his wristwatch, surprised to see his shift was about over. He turned, catching his supervisor's eye and waving him over.

"Sir, I'll be relieved in about thirty-five minutes, but I had a wild idea I'd like to run by you before I bail," Kucera said quietly. The young officer leaned closer. Pointing at a small graph on the computer screen, Kucera explained: "I pulled bus-health data from the last pass of that first bird, the one that wouldn't take my update a while ago. This is the line-level parameter from the onboard KI-57 crypto-box receiver. I've plotted receiver data downlinked over the time we had that bird in Australia's comm window. It doesn't compute, sir."

Kucera moved a stylus, driving an arrow-point cursor back and forth along a perfectly flat line on the touch screen. "So, what am I supposed to see?" the lieutenant asked, a bit sharply.

"We should see some variation in that line; some up and down blips as the receiver draws power," Kucera said, jaws tensing. "See this old data, from the last update a few days ago? It's rippled, because the line voltage fluctuates a tiny bit when the 57 receiver's passing our update-data package. But today, no variation. Voltage stayed rock-steady. I'd say the crypto-box receiver isn't working—but that's a wild-a . . . uh, just a guess, sir."

The lieutenant nodded, then straightened, absently eyeing the computer screen. Kucera liked the officer. He was young and inexperienced, sure, but a damned good engineer who knew his "arcs and sparks."

"Thanks, sergeant. Nice work, too. I'll get our NSA guy on the horn and see what he has to say." The local National Security Agency rep knew this space crypto gear cold, and should be able to quickly determine what was going on with the Navstar's Space Communications Security KI-57.

One thing was for sure, though. Four out of twenty-four GPS birds were putting out increasingly bad timing and positioning data. And all four were fairly close together in the constellation. At times, they could be the only GPS spacecraft in view of a receiver on the ground. Not good for any user who was depending on those birds for accurate timing and position data. Not good at all.

6 APRIL/AIR FORCE SPACE COMMAND (AFSPC) HEADQUARTERS, PETERSON AFB, COLORADO

The chief of USAF Space Command's public affairs office ran a hand through his graying buzz-cut, while he reached for a cup of coffee. Waves of steam were rising off the strong, dark-brown liquid, brewed just the way the colonel liked it. He held the cup to his lips, taking two or three sips while scanning a summary sheet, reflecting absently that AFSPC—and he, in particular—had one hell of a public-perception problem on their hands. Civilians who relied on the complex network of communications satellites were getting pissed. The twenty-first century was the age of instant wireless communication. But, suddenly, communications were no longer "instant."

Americans rapidly were discovering that orbital space had become an integral part of their daily lives. In the past, few citizens had paid attention to small skyward-pointing satellite dishes sprouting from the roof of their local gas station, unaware that the antennas were their end of satellite-communications links to Shell, Chevron, BP, and other gasoline-company computers, where customer credit-card charges were approved in an instant. Because several comsats, the critical space-based link in that credit card–approval lifeline, now drifted lifelessly in the black cold above, their digital transponder circuits fried and worthless, frustrated commuters were consigned to long lines at service station counters. Pay-at-the-pump convenience was suddenly a memory for them. Now, because the machines' space links were inoperative, those drivers had to wait in line at counters to pay six bucks a gallon—thanks to supply problems caused by those damned Iranian hothead clerics. And gas lines were becoming an all-too-familiar sight again.

Down the street, banks' automatic teller machines suffered a similar fate. Unable to communicate via satellite with their home offices' huge computer databases, ATMs refused every bank card fed to them. Consequently, more long lines became the norm, this time at drive-up or teller windows. Banks had reverted to the painstakingly slow methods of the 1950s. Customers' tempers flared.

Once-efficient hospital staffs no longer reached out and virtually grabbed on-the-move physicians by simply paging Dr. Smith or Jones. Millions of doctors' pager/communicators relied on commercial satellites that had gone silent. Harried, overworked nurses and young candy-stripers raced through hallways, poking their heads into patients' rooms, searching for physicians making their rounds. Doctors working in offices a few buildings away or relaxing on a golf course across town dutifully carried advanced communicators that combined cell phones, e-mail communicators, pagers, and myriad other tools in a single, compact handheld device, believing they were but a call away from their patients. Staffs had shrunk over the years, thanks to increasing reliance on instant communications with physicians, and now there simply weren't enough people to send on physician-search missions. Patients suffered, waiting for doctors who no longer responded quickly.

Other medical services began to suffer, as well. By 2010, hospitals had become completely reliant on wireless communication protocols, mostly satellite-driven. When Dr. Jones at Los Angeles Cedars Sinai needed an EKG baseline for one of his patients, the comm link to NYU Medical Center was normally only an access code away. Dr. Jones's computer screen would soon display Dr. Smith's previous EKG report, which could be compared with the heart patient's real-time EKG, obtained by subcutaneous microsensors. But with satellites out of commission, FedEx had to deliver the EKG a day later. And ironically, FedEx drivers normally would have relied on the same communications-satellite infrastructure to transmit their location to the company's central computers. But no more. Now whole systems were crumbling as one satellite after another fell silent.

Similarly, multinational businesses that relied on instant communications to manage far-flung corporate outposts saw transactions slow to a crawl as disrupted transoceanic telephone and data links were rerouted from satellite to terrestrial pathways. Even old-fashioned paper fax messages slowed to a painful inch-by-inch pace, and attempts at voice comm were met with steady beep-beep-beep busy signals. Undersea cables strained to keep up with sharply increased demands, no longer augmented by high-capacity communications spacecraft. A New York–based executive fumed to a reporter, "Doing business these days is like running through molasses! You get nowhere fast!" Indeed, global businesses collectively were suffering billions of dollars in lost revenues, thanks to now worthless, dead communications satellites coasting through space.

Americans on the cusp of the twenty-first century's second decade expected information at their fingertips, available instantly, the AFSPC colonel reflected. They had become accustomed to such convenience, and when services failed, when the world was on the verge of grinding to a stop, people looked for someone to blame. And the folks responsible for keeping satellites in orbit and working were the men and women in uniform. *They* were the ones being blamed.

The entire military space community was on the hot seat, and the AFSPC public affairs chief had more bad news for his boss, the Space Command four-star. The Colorado Springs–based AFSPC headquarters staff was working around the clock, trying to make sense of a deteriorating space situation, while searching desperately for ways to mitigate it. In Nebraska, General Howard Aster and his STRATCOM staff also were scrambling, briefing Pentagon chiefs, Congress, and the National Security Council about growing casualties among commercial, intelligence, and a variety of military satellites, their clipped reports delivered during tightly held secret sessions. Aster's "Tiger Team" of commercial space executives, engineers, and military experts, still huddled at STRATCOM headquarters, had launched a focused wargame aimed at narrowing the list of potential sources of devastating attacks on both government and commercial spacecraft. The most likely source's location appeared to be in the "'Stans" region—perhaps in the still lawless Afghanistan outback—but they hadn't identified the kill mechanism or precise position. Pressure to find and neutralize the attackers was mounting rapidly, and threats of "heads rolling" were being muttered in the halls of Congress.

The public affairs chief, now into a second cup of early-morning coffee, gathered his briefing notes, took a deep breath and headed for the commander's office. *This could get ugly*, he thought with a grimace.

6 APRIL/THE PENTAGON'S "TANK"

Secretary of Defense T. J. Hurlburt glanced at a dozen or so stern faces turned his way, satisfied that all required players were present. Multistarred generals and admirals, flanked by civilian secretaries of the Army, Navy, and Air Force, deputy SecDefs and other senior executive service specialists, ringed a conference table that sported partially submerged computer screens, keyboards, and advanced command-center telephones. The relatively small, highly secure, windowless conference room—dubbed "The Tank" decades ago—was crowded, its periphery lined with aides and staff officers sitting or standing. A large projection screen filling one wall was now dominated by General Howard Aster's bigger-than-life image. Thanks to a secure comm link guaranteeing uninterrupted military video conferencing, Aster and his space experts at Offutt Air Force Base were virtual attendees at the Tank meeting, poised to answer questions raised during the SecDef's emergency brainstorming session in Washington, a thousand miles away.

"Howard, give us an update on the latest. You've lost several more national security birds?" Hurlburt asked, disbelief clearly evident in the tone.

"Not exactly, sir. But we might be in worse shape than if we *had* lost them altogether. Four GPS Navstar spacecraft and one DSP—that's Defense Support Program, a missile early-warning platform—have been hit. All five are still working, to a degree, but not like they should be. And the combined fallout of their degradation is what's got us damned worried. If I may . . ."

"Yeah, go ahead," Hurlburt interrupted. "I want everybody on the same page, so give us a rundown on what the hell we're up against here."

A retired Army general who had made a highly unusual U-turn back to the Pentagon—this time as a civilian political appointee—Hurlburt was legendary for taking trouble head on. He was not a beat-around-the-bush guy. Rather, avoiding even the pretense of political maneuvering, he was just as intolerant of posturing and what he called "polspeak" in his current role as he'd been during the mid-2000s, when he ran U.S. Central Command as its four-star commander. Nicknamed "Bull" when he was at West Point, Hurlburt was credited as the brains behind the successful combat campaign in Afghanistan that had finally decimated al Qaeda and the Taliban. His hard-hitting preemptive strategies had made impressive dents in global terrorism. The old soldier had completed his military career on a wave of official accolades and honors bestowed by Washington, then retired to a slower pace in North Carolina as a gentleman farmer, watching soybeans grow and fresh paint dry. For the first time in their forty-year marriage, he and his wife had actually enjoyed their days, weeks, and months, free of Army disruptions and the midnight "they're-firing-on-Fort-Sumter" phone calls. But it was a short-lived breather.

Always intellectually restless, Hurlburt had soon conceded to leading a few studies of ongoing Pentagon "transformation" initiatives. In 2007, he also had briefly advised a long-shot presidential candidate, translating cryptic Pentagonspeak into straightforward, twenty-first-century defense policies the electorate could understand. Overall, he had managed to maintain a relatively low profile, working behind the scenes. When the brutal, highly divisive 2008 presidential elections were over, however, he'd answered the phone one morning to hear the new commander in chief offer Hurlburt the SecDef job. That long-shot candidate he'd advised was now the president-elect and needed a trusted voice speaking from the Defense Department's largest E-ring office. Ever the supportive, yet reluctant trooper, Hurlburt's wife had sighed and started packing their experimental fuel-cell Chevy Suburban for yet another move to yet another home near the Potomac.

That relaxing farm, that phone call, that trip to Washington—all were distant history now as Hurlburt stared at Aster's image floating on the wall-screen. Aster, in rapid-fire terms, aided by stark PowerPoint graphics, delivered the bad news. The nonresponsive DSP satellite had left a hole in America's missile defense network, but the remaining DSPs and newer Space-Based Infrared-High platforms—still in their checkout phase—could compensate. His biggest concern was how an incredibly robust, well-protected, Cold War–era spacecraft could possibly have been attacked successfully. It was 22,500 miles from Earth, and hardened to withstand both powerful cosmic rays and the gamma- and neutron-flux of an in-space nuclear blast.

Four GPS navigation satellites were still operating, but chronic errors that naturally crept into all GPS satellite position and timing circuits could no longer be canceled by 2SOPS operators at Schriever AFB. The spacecraft wouldn't accept update packages that normally would have compensated for those natural errors. Apparently, the front-end preamplifiers of new-generation National Security Agency KI-57 cryptographic receivers on the mid-Earth-orbit navigation spacecraft had been disabled by an unknown attacker using an unknown weapon. Probably the same party that had taken down several civilian satellites, Aster's experts surmised.

Unless these units received the right access codes, at the right time, NSA KI-57s riding on all milsats were designed to block any commands beamed from the ground, logically concluding that the signals were unauthorized attempts to take control of the birds. In essence, the boxes were NSA's electronic gate guards, charged with making sure bad guy–transmitted command signals did *not* reach the sensitive hearts of GPS and other satellites. But now-inoperative preamps had cut the lifelines to four GPS satellites' microcircuit innards, leaving the spacecraft free to spew increasingly incorrect timing and position signals to unaware terrestrial receivers.

"In other words, those two GPS birds are putting out bogus information, and we can't get error corrections into them," Aster summarized. "Every update-load we transmit is completely blocked by those fried front-end KI-57 circuits. There's no way to get through the damned NSA boxes, so we're really no longer in command of the spacecraft. Our guys are sending updates on schedule, just like they've done for years, but the satellites won't accept them," Aster restated, pointing to simplistic block diagrams on a slide.

"That's actually worse than if the birds had died completely. Little by little, a sizable chunk of our entire GPS constellation is degrading, because four satellites' signals are unreliable. Initially, we didn't think this would present much of a problem, because most GPS receivers on the ground and in aircraft can ignore bad GPS data, at least from a few birds. The receivers look at all satellites in view at a given time, take a vote, and throw out any whose data deviate too far from an average."

"So, what's the problem, then?" Hurlburt asked. He knew the sickening answer, but some around the table didn't.

"Sir, what concerns us is this: The SOB who's attacking our spacecraft—and with amazing success, I have to say—if he knocks out the crypto-box front-ends of several more GPS birds, the whole thirty-spacecraft 'ball'—including our on-orbit spare birds—will degrade rapidly," Aster added. "In other words, we won't be able to trust GPS signals from the entire constellation, because it'll gradually become less and less accurate. Terrestrial receivers will blindly vote-in bad data, because the bulk of information coming from degraded spacecraft will be bad to start with. We've already had one tragedy, simply because those four 'bad' satellites happened to be the only birds in view when we had a covert strike under way last night. An incredibly low-probability fluke, but it happened."

Hurlburt whipped a glance at the Special Operations Command four-star, a ramrod-straight Marine Corps general seated to the SecDef's right. The marine's left fist, resting on the table's polished surface, tightened noticeably.

"Howard, all of us, at least those cleared for that information, know what happened in the desert last night. You're already attributing that . . . that screw-up to degraded GPS signals?" Hurlburt asked incredulously. Being notified only hours ago that four American SOF troops had been blown to hell by American bombs had shaken every man and woman in the Tank.

"Absolutely, sir. We're positive that gross GPS errors caused that incident. A freaky combination of factors, yes, but the outcome was deadly."

"Good Lord! And we could be facing even bigger disasters," Hurlburt groused, shaking his head slowly. "Airliners all over the world use those GPS signals to navigate, and most now rely on them during landing approaches, right?"

"Gets worse, Mr. Secretary," Aster continued, lips tightening to a thin line. "On the military side, our bombers and fighters could have one hell of a time just finding targets, since they all use GPS for primary navigation. Damned few of today's pilots know what 'time-and-distance' or 'dead-reckoning' navigation means, because they haven't had to do it in years. Oh, they can fall back on their inertial navigation systems all right, but pilots, for at least a decade now, have depended on GPS to automatically update those inertial nav systems. We can deal with that, though. Not real fast, but we *can* get workarounds in place. Unfortunately, two F-22 pilots last night opted to believe their GPS numbers instead of the INS coordinates. That's how they were trained to respond to a GPS/INS discrepancy, though, so they're not to be blamed.

"As they discovered, another critical impact is on our precision weapons," the STRATCOM chief continued. "Most of today's U.S. air-to-ground arsenal depends on strap-on GPS kits for terminal guidance. Some have INS backup systems built in, and our newest weapons also can be laser-guided. The wingman on last night's mission reverted to a radar-guided E-JDAM attack to take out the . . . uh . . . the target, even though that option can be less accurate than GPS guidance. But, in general, we've come to rely heavily on GPS as our primary means of all-weather targeting. For the moment, those precision-strike capabilities are definitely degraded, at least in some parts of the world. And those 'parts' will spread, if more of our GPS birds take hits."

"What about the ICBM and SLBM fleets? Are they in the same boat?" Hurlburt asked, not sure of the answer, but afraid he wouldn't like it. Intercontinental ballistic missiles and their submarine-launched counterparts were still the core of America's nuclear posture. If degradation of the GPS constellation reduced the nation's nuclear strike options . . . *Shit! We could be in a massive mess!* Hurlburt realized.

"That's not a major issue at the moment, Mr. Secretary. All of our missiles still rely on a very sophisticated inertial nav system, with GPS used only as a cross-check. We can tell the nuke birds to . . . well, basically to ignore the GPS signals altogether," Aster summarized. *No sense in losing these folks in nitty-gritty details,* he thought.

Hurlburt sensed that the Tank's key players had grasped the gist of what STRATCOM was dealing with, but maybe not more-subtle ramifications. "So, what's the bottom line, Howard? What do I need to be sweating here?"

Aster stared into his screen, the built-in high-definition digital camera picking up every line in his face, as well as the cold blue eyes that seemed to bore into every person in the Tank. "Mr. Secretary, we're approaching a situation where the United States is partially blind, deaf, and mute in a space sense. If we lose a significant percentage of our overhead intelligence-gathering capabilities—imaging and signal-snooping, if you will—on top of the one missile-warning bird, four GPS platforms, and several critical commercial satellites already lost or in trouble, we'll be in a major national security hurt. Already, we're having difficulty communicating with some of our critical intel and military assets across the globe, at least in a timely manner. This means that if a bad actor, or several bad actors, want to make mischief, this is the time to do it. We simply aren't able to catch 'em in the act right now."

"So, let's clamp down on every tidbit about this situation," suggested a civilian defense undersecretary. "We make this compartmented information—Top Secret-Special Access Required, as of right now."

"Sir, it's too late for that," Aster answered, grim-faced. "An hour ago, I fielded a call from an *Aviation Week* reporter, who not only knew exactly—*exactly!*—which commercial birds we've lost, but had a pretty good handle on the ramifications of those losses. He asked us to confirm that all government spacecraft were still operational. We told him we couldn't comment at this time, but I think he's already gotten a whiff of the GPS degradation. While TV and newspaper folks were screaming about gas pumps, doctors' pagers, and ATMs, these *AvWeek* guys were talking to someone on the inside. Or maybe they just figured it out, I don't know. But they're ready to go with a story detailing how badly we're hurting. . . ."

"Jesus!" Hurlburt interrupted. "Didn't you tell the sumbitch to go pound sand? We can't admit to what shape we're in, or every nut case in the Middle East and Pacific will jump on our butts!"

"I've asked them to sit on the story for a while, sir. Their editor-in-chief agreed to, but I don't think he can for long," Aster clipped. "It's just a matter of time until the TV animals get wind of this, and they'll really screw the story up. I'd rather have *AvWeek*'s guys do their number on it first. At least they have a reputation for getting the story right—and they're willing to work with us. Might take your intervention, though, Mr. Secretary."

"All right, all right. We'll work that one off-line, Howard. Anything else we need to know?" Hurlburt asked, anxious to end the stream of depressing news. Aster quickly outlined the wargame underway at STRATCOM headquarters, aimed at locating the source of attacks on U.S. spacecraft, identifying the weapon or weapons being used, understanding why this nightmare was happening and, finally, what its national-security ramifications might be. Then the general leaned forward, eyes hardening as he stared into the videoconferencing screen.

"Sir, I need *immediate* access to a super-secret spaceplane that *someone* has developed, but never bothered to brief me, the nation's designated *space* combatant commander, about!" Aster spat, face visibly flushed. The general was pissed.

Hurlburt was stunned, Aster noticed. *Aw shit! The SecDef hasn't been briefed either? How the hell can that be?*

"Howard, *what* in God's name are you talking about?" Hurlburt thundered.

"Mr. Secretary, a couple of my 'iron majors' were in here less than an hour ago, saying, 'Sir, we think you need to know about a two-stage-to-orbit spaceplane project. It could help us get out of this mess,' or something along that line. It seems our deep-black spook friends developed a limited-capability system about twenty years ago that could launch a small spaceplane into low-Earth orbit. Two versions were built, one unmanned, the other piloted by a single astronaut. Both types were carried to altitude by—now get this—a mothership that looks a hell of a lot like the old XB-70 Valkyrie bomber."

"But there's only one of those in existence, Howard! And it's in *your* Air Force museum at Wright-Patterson! The only other one ever built crashed at Edwards damned near—what, forty-five years ago? You know that!" Hurlburt's face was darkening, too.

"I thought so, too, sir. The majors tell me a third—and evidently a fourth that later crashed—were built in the late eighties, using a few long-lead items manufactured before the program was canceled after Ship One crashed. Point is, there's at least one flying now; has been for years. Supposedly it can carry a small spaceplane to high altitudes and supersonic speeds, then drop the little bird, which fires its aerospike engines and zooms into space. It can carry a microsat or a variety of recon packages, I'm told. It was supposed to be an incredibly effective surprise-recce platform, because it can be launched in any inclination."

Hurlburt scanned those around the table. Most of the faces mirrored his confusion. Only the National Reconnaissance Office director looked uncomfortable.

"Damn it, Marty! What's the story here?"

"Mr. Secretary, I'd prefer to take this up privately. We're into an area that's. . . ." Martin Timm, a slight, bespectacled career bureaucrat turned political appointee, muttered, hemming and hawing about the "need to know."

"Like hell we will!" Hurlburt interrupted, temper flaring as it had in his old, general-officer days. "Just how long were you boys going to sit on this? Until the whole goddamn satellite network melted down and we were pissing all over ourselves trying to find a goddamn target?" the SecDef bellowed. "We have a very serious situation on our hands with American casualties already taken in the field. I don't give a flyin' flip about stupid, secret, spook handshakes and black-world, off-the-books science projects! Christamighty! How can I—or a combatant commander like Howard—run a damned railroad if we don't even know what trains are out there? If you have this spaceplane thing locked up in a barn out at Groom Lake or somewhere else, and it can help Howard get a handle on this space fiasco, I want you—*you, Marty*—to put it at his disposal *right now!* This minute! Is that clear?" Hurlburt pounded a fist on the table, underscoring his verbal blast.

Timm's Adam's apple bobbed as he visibly cringed, nodding vigorously. "That I'll do, Mr. Secretary. But it's not quite that simple." He turned toward the big screen, still dominated by Aster's now-red forehead, accentuated by the border of snow-white hair. "Howard, the Blackstar system hasn't flown in at least ten years. Frankly, it didn't work out as well as we'd hoped it would, budgets got tight and the program was shelved in the late nineties." Sensing an imminent second explosion from Hurlburt, Timm rushed ahead.

"That said, a mothership and one spaceplane have been maintained in airworthy storage at a remote, highly classified location, just in case. They can be ready to fly in about a week, I think, but rounding up flight crews and getting them checked out again will be the pacing factor. As a first step, I'll get a team out to Omaha today. They'll brief you on what the system can and can't do, Howard."

Aster nodded and brusquely thanked Timm. From halfway across the nation, the STRATCOM chief locked eyes momentarily with the SecDef. In the Tank, the ex-Army four-star nodded back. *The friggin' spooks have done it again,* his hard eyes seemed to say. Aster silently agreed.

3

WARGAME'S NO GAME

*Those who are skilled in combat do not become angered;
those who are skilled at winning do not become afraid.
Thus the wise win before they fight, while the ignorant fight
to win.*

Sun Tzu
Sign over entrance to STRATCOM Wargaming Center

7 APRIL/SECRETARY OF DEFENSE RESIDENCE, WASHINGTON, D.C.

Gently swirling a long-stemmed glass of Pinot Noir, aerating
it before sipping, T. J. Hurlburt settled into a well-worn
leather chair and flipped through *Aviation Week & Space
Technology* magazine's newest issue. It had been hand-
delivered to his office—as it always was—only hours after
publication. As he'd left the office that evening, though, a
stern military aide had given Hurlburt the magazine, suggesting the SecDef take a look at the lead story "at your earliest
convenience, sir." The colonel knew Hurlburt rarely read
Aviation Week or any other publication, preferring the digested "Early Bird" summary of pertinent news the Pentagon
staff assembled in the wee hours every morning.

"What's so special about this thing, Colonel?" Hurlburt

had asked, grabbing the publication without breaking stride. He had been late for a dinner party his wife insisted they attend, and was racing the clock.

"It's the space situation, sir. Those satellite losses. *AvWeek*'s opened a can of worms," the aide had clipped, matching Hurlburt's rapid pace.

"Can't it wait? I've already got a boatload of garbage to read tonight," the SecDef had groused, racing down an echoing stairwell. The old warhorse hated waiting for elevators.

The colonel had hesitated, matching Hurlburt's pace down the stairs. "For a bit, Mr. Secretary. But I'd highly recommend you take a look . . . well, before you hit the sack tonight."

Hurlburt had thrown him a dark glance. "Oh? And exactly why is this so damned important, Dirk?" Everybody on the Pentagon's E-ring knew very well the SecDef was absolutely anal about getting his six hours of sleep, no matter what crisis the five-sided building was dealing with. Hurlburt would be awake long before sunup, usually arriving in the office before 5:30 A.M. But nobody—*nobody*—screwed with his bedtime.

The colonel had pushed open a heavy outside door, noting that the SecDef's limousine was waiting. "Sir, because I suspect the president will want to talk about this at oh-dark-thirty tomorrow. If not before."

Hurlburt had stopped quickly, turned and held up the now-rolled magazine. "Because of *this?*" The colonel nodded. With a grunt of amazement, the SecDef had bounded down the steps toward his car.

Now, alone in the quiet of his home office, Hurlburt tossed a necktie aside. He scanned the *Aviation Week* article quickly, then slammed the publication onto a small table beside the chair. "Shit!" he blurted out. Standing abruptly, he walked to a window and stared into the darkness, sipping the wine while his mind raced. Minutes passed before he returned and dropped into the chair again. Hurlburt retrieved the magazine, turned to a blue-tinted sidebar, on a page adjacent to the primary, fact-filled news story about satellite losses, and reread the last few paragraphs of a section marked "Commentary":

This recent surge of spacecraft losses has far-reaching ramifications that should not be underestimated. A Pentagon official familiar with the as yet unraveling situation admitted that the nation is at least partially "blind, deaf, and mute" in a space sense, because imaging, communications, missile-warning, and signals-intelligence platforms have been adversely affected. To what degree is difficult to ascertain. However, a high-level emergency session in the Pentagon's "Tank" apparently caused considerable angst among senior military and intelligence officials.

One might add "lost" to that blind-deaf-mute list, as well. Unconfirmed reports suggest that the Air Force is unable to communicate with several satellites in the Global Positioning System constellation, which provides highly accurate timing and positioning signals used by millions of people, from airline pilots to fishermen. That could lead to ever-growing inaccuracies, because critical updates of those satellites' payloads would be impossible, according to a GPS-system consultant. For combat forces, such inaccuracies could have a major impact, forcing aircrews to rely on inertial navigation systems rather than their ubiquitous GPS sets. Depending on the extent of GPS inaccuracies, U.S. precision weapons could be affected, as well.

But the most critical questions now facing national leaders must be: What are we missing out there? What are terrorists and potential adversaries up to? How have these spacecraft losses affected our ability to command and control military forces in the field, which rely on robust "reachback" links to carry out far-flung operations?

Without question, U.S. space situational awareness has suffered from the loss of commercial imaging platforms, and, reportedly, degradation of national security space assets. If the Pentagon and intelligence community have lost space-based capabilities to monitor—visually or via

signals interception—the activities of rogue states and ter-
rorist cells, does that not put the nation's citizens at signifi-
cant risk? If an adversary is bent on creating havoc for the
United States and its allies, what form might that take in
the absence of around-the-clock surveillance? More impor-
tantly, could the precursors of an attack against United
States and allied interests be detected soon enough to pre-
vent or mitigate such an attack?

Such concerns are just some of what appear to be very
serious issues now under discussion in Washington and the
headquarters of regional combatant commanders. And
unless the causes behind these satellite losses are determined
and "cured" very soon, one must assume that the U.S.
security posture will deteriorate rapidly. This situation
already has triggered increased global instability, and a
shooting war is not out of the question, according to the
same Pentagon official.

"This whole mess could have been avoided, had the
nation's space resources been better protected," he said. . . .

Hurlburt reached for the secure-line phone at his right
elbow and punched a speed-dial number. General Howard
Aster would be losing sleep tonight, too.

8 APRIL/STRATCOM WARGAMING CENTER

Colonel Jim Androsin pored over a series of data sets representing all known engineering parameters related to satellite
anomalies over the past several years. He raised his head
every few minutes to survey Analytical Graphics' Satellite
Tool Kit (STK) visualization of spacecraft failures and their
terrestrial correlation, searching for a potential origin of interference. Androsin's team had already assembled a Segment
Impact Vulnerability Matrix—or SIVTRIX as it was known
by gamers—to help narrow the possibilities, and he was
looking for insights. Something—anything—to help understand the unfolding crisis.

Androsin framed a question to aid his own reasoning. *Okay, Jimbo, we're getting to the what, but who, and why?*

SIVTRIX had been developed in the late 1990s as the United States became increasingly concerned about space assets protection. The system took into account three aspects of the entire U.S. space infrastructure: the satellite segment, or "ball" as space geeks referred to man-made space constellations; the control segment, comprising ground stations, control facilities to monitor and fly the balls, plus sensors for keeping track of the ball's health; and the user segment, a distributed architecture that provided downlink signals to any end user, be it an F-22 cockpit or freight trucks along America's roadways. Androsin needed SIVTRIX data loaded immediately. There was much at stake, and time was clearly on the side of an adversary.

"Jill, where are the nodal analysis inputs for the vulnerability matrix?" Androsin called.

"Workin' it. Waiting for some additional data on the GPS ball." Jill Bock was one of Androsin's most trusted decision-support tool experts, and was fast becoming a world-class wargamer in her own right. Not only did she have an extraordinary understanding of technologies adapted to the wargaming center's decision-support tool, she was also a wizard at categorizing campaign-planning criteria for the many inputs the modeling and calculation process called for. Jill was a rarity among her generation: a young mathematical genius who believed the abstraction of numbers held an eternal quality. Nevertheless, she'd argue passionately that the human mind could essentially outperform the best of computer models. Such open-minded thinking had brought her to the fore of classic wargaming practice.

"Come on Jill, we've got to get the analysis loaded—now!" barked Androsin impatiently.

"On its way, boss." *Man, sure are some unusual anomalies with GPS,* she thought. "And I want to be sure we're still tracking closely with the general's objectives."

Jill's whiskey-soaked voice kept a listener enthralled with its melody, and her striking good looks were framed by an eye-catching cascade of brunette hair. There was a distracting quality to her presence.

"Right. Okay, let's review the objectives one more time, before we start the game," the Army colonel suggested, dragging a chair close to her workstation. Androsin felt a growing unease as he reviewed intelligence data Jill was assembling, her fingers flying across the keyboard.

Finding out who's behind these satellite outages will be damned hard. And the spooks don't have a clue, he mused, recalling General Aster's lament during the previous night's teleconference. The general had commented that America's national security was in jeopardy as a result of spiraling space-system losses. And no one knew the cause. *Right now, there just aren't any good options to recommend to SecDef and the president,* Aster had summarized. *When we develop some, the president will want us to answer why. Why is this happening? And who's doing it? We can't just produce a laundry list of what went wrong on what systems. What he really wants to know is whose ass are we gonna kick. . . .*

Jill sensed Androsin's tension. She'd rarely seen this Renaissance warrior visibly nervous. Androsin's tours of duty in the crucible of change throughout the Middle East and Southwest Asia had created a keen understanding of those profound elements that stirred winds of change. He had proven to be a natural diplomat, synchronizing U.S. and coalition forces with international organizations to suppress provincial infighting from Afghanistan to Iraq to Uzbekistan. And, as a seasoned warrior, he had skillfully enforced security in the provinces, ensuring stability needed to nurture the growth of fledgling democratic practices. His troops knew him as a wise leader infused with humanist tendencies. And his adversaries quickly suffered his fierceness when violence ensued.

Jill, on the other hand, had graduated from MIT with a Master's in quantum physics, yet remained enthralled with the wonder of abstract mathematics. Still, she'd never dreamed that she would wind up tricking information out of a supercomputer buried inside a digital war room filled with wall-to-wall ex-fighter pilots, Special Forces types, and spooks. Jill had met then-Major Androsin during her first critical national defense job, a project developing a national security and operational decision-support tool. Androsin had a track record of developing the battlefield utility of complexity theory, and her confidence in her own abilities had grown rapidly under his tutelage.

"Colonel, what do you suppose is really happening here?" she asked, returning to the task at hand. "I mean, none of this seems to track with what we're used to seeing."

Androsin didn't answer immediately. He couldn't shake that sense of uncertainty in General Aster's tense, frustrated tone the night before: *Jim, we're at war, dammit! I feel it, but I can't see it yet. Get those wargaming teams in gear and get me some answers!*

"That's why we're wargaming," the colonel replied. "Sorting through the magnitude of uncertainties. Examining our own capabilities and bouncing them against a range of desired outcomes. This is all about introspection, about looking inside our beliefs, our understanding of adversaries, and testing concepts for action."

Wargaming was Androsin's passion. He fervently believed in the process of "gaming," even though the term had an unfortunate connotation, given the proliferation of computer-based entertainment. Although he kept a healthy sense of cynicism about wargaming's value in certain applications, he fervently argued that strategic and operational wargames were essential components of a nation's critical thinking.

"We'll need to feed our team leaders everything, as we get it," he added. "No time for analysis and distillation."

"Sure, boss. What about the satellite data? Could take a while to get the ninety percent solution . . ."

"Another reason to press on with the wargame," Androsin responded, somewhat patronizingly. "Avoid the analysis-paralysis syndrome. There's never enough information in the real world for a ninety percent solution, but at least gaming lets you test the consequences of decision making with the best info available, while the tyranny of time marches right through your foxhole."

Androsin had studied the extensive history of wargaming, which extended from Sun Tzu, the brilliant Chinese military strategist/philosopher, through the nineteenth-century Prussians' scaled versions of war known as *Kriegsspiel*. He came to admire the U.S. Navy's refinement of the art during the twentieth century—particularly in the runup to World War II— and its Maritime Strategy during the Cold War that followed. There was no question in his mind about the value of socializing tough issues, and creating a no-harm environment, a safe laboratory of sorts, where the gamut of tactical, operational, and strategic decision making was examined along a spectrum of alternative circumstances and futures. *Wargaming* was all about challenging one's own thinking, a complex way to keep from drinking your own bathwater, as he was known to say.

Androsin embraced an ancient Taoist view, simultaneously appreciating the many dynamics affecting human behavior, but on a grand scale. His master's thesis for the Naval War College's National Security and Strategic Studies program had involved a method for employing Taoist thinking and understanding the many dynamics that shaped nation-states, even rogue actors. Influential people in and out of government believed his thesis was a practical roadmap for resolving contradiction and paradox in international relations. In short, wargaming methodologies were a key element in the process of gaining knowledge, then framing solutions.

"Whoa, Nelly! More bad news, boss." Jill pointed to the latest stream of information scrolling across her intel screen. "Chicom's are maneuvering some nanosats?"

Androsin reacted. "We've gotta get this wargame in gear! Too much happening too fast. Load the SIVTRIX, just as it is now. And have your techs stay in manual feed during the vulnerability assessments."

The colonel was now in the vanguard of a charge to bring wargaming out of a social paradigm—"grip-and-grin" was the derisive term he used—into a world of examining the interpersonal and global aspects of conflict. He embraced breakthrough technologies, such as artificial intelligence, complexity theory, and logic derivatives, integrating them with the expertise of learned people to create large expert knowledge databases for decision support. As his thesis had proposed long ago, Androsin had recently argued successfully for a cadre of academic, diplomatic, economic, social, and military experts to partake in a series of operational and strategic wargames. Their charge now was to examine new ways for dealing with a chaotic and uncertain world.

This confluence of technologies and expertise had created a resurgence of critical thinking and decision making that had atrophied in the 1990s and 2000s. Here at STRATCOM, Androsin and Jill were at the nexus of this revolution, and they knew it. They were the choreographers of an interactive process that created unexpected, yet insightful results, based on a regimen of meticulous analysis. Wargaming was rapidly becoming instrumental in sorting through the "fog of peace" in much the same way battlefield commanders dealt with the "fog of war."

Jill used a voice-activation device to access the imminent wargame objectives, displaying them on her computer's primary screen. Androsin then read the objectives aloud, willing greater clarity into the mounting uncertainties and impending calamity.

"Examine all satellite anomalies to date," he ordered the system. Its natural-language voice processor quietly parsed sentences and generated object code, bringing to life an array of deeply stacked parallel processing circuits. "Assess patterns of anomalies. Determine likely causes and linkages to what and who is behind any deliberate interference. Project likely risks for space and terrestrial systems, critical infrastructure, U.S. and allied forces and population centers. Explore options for countering space system interference and . . ."

"Why don't we just call them *attacks?*" Jill interjected somewhat testily.

"Hang on, Jill. We don't know . . ."

"Oh sure! If they were airplanes or ships, and people were dying, I'll bet the urgency would be different. These damn satellites just don't have mothers!"

Androsin suppressed a smile, amused at Jill's outburst, then continued his voice commands to the computer. "And last, *explore* options and *recommend* courses of action to include terrestrial responses that allow for a gradual escalation of force to accompany diplomatic demands."

Androsin hesitated, then sighed before adding, "Use back-channel systems to develop and propose massive retaliatory responses for higher authority's consideration in discharging duties necessary to ensure the survival of the United States of America." He gave Jill a knowing look, deselected the voice-activation system, and returned to his own workstation.

Swinging into a mode of increased action and intensity, Jill manipulated an array of screen-activated buttons, commanded voice actuators, and barked directions to technicians huddled in front of their individual stations across the room. The scene had become a cross between a submarine's control center and the imaginary set of a starship. The wargaming battle lab was home to a systems architecture for the "BOYDTRIX" decision-support tool. The acronym had been coined by acolytes of USAF Colonel John Boyd, who had developed critical ways of thinking through complex problems. He also was the father of the "OODA Loop," the Observe, Orient, Decide, and Act mantra of fighter pilots, who strived to get "inside their opponents' decision loop" as a way to ensure victory in aerial combat.

The OODA Loop would come to have much broader applications, however, because many of the corollaries first developed by Boyd would later be refined by the pioneers of complexity theory. Jill, with a background in abstract mathematics and game theory, had quickly become an expert in the field and a key contributor to developing the BOYDTRIX system's computation and 3D display characteristics. Today, this decision-support tool engulfed two stories of the battle lab and more than 75,000 square feet of floor space.

Wary of the way men create hidden-meaning nicknames, Jill believed the phonetic sound of BOYDTRIX *must* have a sexual connotation. But she never asked.

"The new GPS data from Schriever is gonna play havoc with this run," Jill declared.

"Why? What's the problem?" Androsin asked, eyes still locked on his workstation's plasma screen.

Jill took a deep breath, dropped several data sheets on her desk, and walked deliberately to Androsin's station. "The space geeks are saying anomalies on four of the GPS birds are related to crypto gear," she said, hands on hips in a defiant pose. "And my initial BOYD query says there are no known probabilities for such failures."

"Come on, Jill. It's not like you to say 'no known prob—'"

"I should say no *programmed* logic, Colonel," she interrupted with impatience. The lady could be damned impertinent, Androsin noted.

"Yeah, whatever. Look, the boss needs answers fast. Let's move. We'll brief in the ops center." Androsin turned to the duty communications technician. "Call the Red team members in, and let Admiral Lee know he has the conn."

Opening a door, he winked at his gaming partner. "Jill, hang on. I think we're in for a doozy."

8 APRIL/SOMEWHERE IN SOUTHWEST ASIA

A slight, stooped figure fiddled with a complex array of thick cables, his fingers tracing one's path to its terminus. So engrossed, he was oblivious to a thickset man who watched from the doorway. The latter sniffed, detecting a hint of ozone in the dank air. He wondered again how the Russian scientist could tolerate hours in this cold, damp basement filled with heavy, paint-scarred electrical equipment. Amazingly, the jumbled mess worked, though.

"Hola, Lexi! What de hell you doing?"

The older, thin man jumped, startled. He scowled, but quickly managed a weak smile before turning toward the rough voice. Alexi felt the bile of derision for his crude benefactor rise in his throat, but swallowed it just as quickly. Domingo's money continued to flow freely, and that allowed the Russian and his family to live far better than they had at Arzamas 16, the closed-city, Soviet-era nuclear weapon design institute and laboratory that had been their home for decades. He had literally disappeared from the lab one day, lured by the mysterious Colombian's intriguing questions—and, of course, the money. In his head, he had carried with him the secrets of an almost completed weapon system that, in its own way, would prove just as influential as a nuclear device.

"Señor Domingo. I am surprised to see you. I thought you had returned to . . . mmmm . . . your 'family,'" the Russian said, his practiced, precise English almost devoid of Russian accent. He knew the short, round Latin had a penchant for wine and women, mixing the two every chance he got, which usually kept him away for days at a time.

"Ah, yes. But I bring very big idea. You want hear?" Domingo asked. His toothy, ultra-white grin, framed by a handsome, smooth-skinned olive face, made Alexi's spine tingle. *A serpent's smile,* the Russian thought. He'd seen Domingo summarily dispatch a snoopy neighbor, the white-handled knife but a flash in the darkness. And he had smiled that way then, even while carefully cleaning the blade on that old man's tattered sweater. Alexi shivered again, thinking of the body, now encased in concrete several feet below the powerful new generator right there, in the basement's far corner.

"Of course. Shall we move to the office? The chill here will be bad for you." Alexi reveled in patronizing the Colombian. The brute never seemed to notice, either. Alexi knew his own intellect far exceeded that of his patron, as he had proven time and again, especially in the last few weeks. The system, guided by Alexi's incredibly accurate long-range ultrawideband tracking radar, had exceeded even his private expectations of its ability to locate and isolate target satellites. His colleagues from Arzamas 16 would be amazed, truly amazed, at the profound results achieved so quickly, he reflected. It was a weapon that would have turned the tide of the Cold War, had the Cold War continued.

With cups of dark coffee between them—a priceless grade-A Colombian roast that Domingo had managed to smuggle through customs—Alexi stole a glance across the table, taking note of Domingo's dark and lifeless eyes. Their contrast with the man's disarming, perpetual smile was unnerving. The Russian, again, felt a telltale tingle at the base of his neck, the same one he'd known years ago, when GRU agents periodically grilled him about his work at the lab. *This being is very dangerous,* that tingle whispered now.

"Your idea, Mr. Domingo?"

"Ah, you very, very good scientist, Lexi. My people, they very happy," the Colombian said, one hand waving. Always the active hands, Alexi noted. Domingo was constantly in motion, especially when his tiny brain was sparking. "De sat'lites; all no work! Good! Good! Comsats, GPS, all dead, yes? *Fantastico!*" He thrust both hands upward, signaling victory, but also opening his coat and briefly exposing a shoulder-holstered handgun. Alexi cringed inwardly, but nodded, dropped his eyes and smiled slightly, feigning humility.

Domingo pulled a document from the flashy suit's inside breast pocket and spread it on the table. In broken, accented terms, he outlined his big idea. Alexi was both intrigued and scared witless. This *was* big! Maybe too big for him. But he dared not show the gut-level doubt that threatened to loosen his bladder. The thought of Domingo's flashing knife—and that gun . . .

"But of course, my friend! A little time, some changes, but I am certain this is very possible." The Russian scientist flicked a hand dismissively. "I have access to the American's 'space catalog,' you know. It is but a matter of finding the right target. A big sky, yes. But I will find what you seek, then . . ." *Smack!* Alexi slammed a fist into his open left palm with surprising force. "The capitalist spy satellite and outpost are gone. Kaput, as our German friends will say!" He immediately regretted the "capitalist" comment. Domingo and his friends fancied themselves as the ultimate capitalists, engaged in multilevel marketing of contraband narcotics: distribution at its unfettered best.

Domingo's cold stare caught the flicker of fear cross the

scientist's eyes before they dropped again. The Colombian slowly leaned back in his creaky metal chair, both palms resting on the table. Fear. He was a master of instilling fear, and he could smell it the way a rattlesnake sensed heat. Alexi was afraid. So satisfying, that power to frighten the weak. The pompous Russian underestimated Domingo, both his intelligence and his understanding of advanced technology. It was better that way, though. Playing the dumb playboy could be very advantageous. Still, he was extremely impressed by the Russian's abilities and the successes he had achieved in such a brief period.

"Ah, yes, Lexi. But we like de money, yes? You capitalist, too!" Domingo laughed heartily, slapping the tabletop. Alexi smiled weakly.

"Mr. Domingo, I may need to . . . mmm . . . procure a few items to accomplish your brilliant idea. That could be difficult . . ."

"Ah, yes, yes! What you need, you ask, okay?" Another coat pocket produced a thick wad of green bills, all U.S. currency, or so it seemed, and in large denominations. From the corner of his eye, Domingo took note of Alexi's tongue flicking across his pale lips. *We like the money, don't we, Doctor? These Russkies always like the money. And they'll do anything for it. Absolutely anything.*

"And, my apologies," Alexi continued, carefully folding the bills. "My daughter has been ill, but she's getting much better . . ."

"Yes, yes, Lexi. I know, I know. She very fine eng'neer. She must help you, I understand. Others, too?"

"Perhaps, my friend. I must think more about your 'big idea.' If it becomes necessary, I will draw on the same trusted associates as before. I prefer to rely on only my daughter, if possible. Much safer, you know?"

Domingo quickly dismissed Alexi, stepped outside the shabby building's only door, and lit a cigarette, his sunglass-shaded eyes sweeping the street. Nothing moved. He slowly walked toward a BMW sedan, the vehicle clearly out of place parked alongside the muddy, potholed road. As he opened the car's door, he glanced upward at a large fiberglass structure extending above the roof of Alexi's innocuous little house. *Who the hell would ever guess?* He shook his head, marveling again at the absolute genius of his brother's plan. And it was working. The dead and dying satellites had already generated far more cash—and at much less risk—than dozens of their cartel's normal shipments.

And his side-deal with the cold, deadly Iranian, a little extra piece of the action, had been his own idea. That deal was paying off well—all in euros, too, a twist that made him even happier, because it seemed to screw the Americans yet again. Domingo turned the ignition key and put the car in gear, reflecting briefly on the new Swiss account he'd opened the previous week. That one was his and his alone.

Weaving the BMW through shabby back streets, Domingo mentally toyed again with his brother's new plan. If Alexi could pull this one off, strike before the Americans even realized what was happening to them, revenues would double or triple. And the oil-rich Iranians would stand toe-to-toe with the arrogant and soon-to-be blinded American bastards, now thrashing around like a blind beast. That was something for which Iran would pay more than a king's ransom. And for Domingo, betting on the supremacy of the euro after the imminent collapse of the American dollar, it was a stroke of brilliance even his smarter older brother could not have dreamed up.

8 APRIL/STRATCOM HEADQUARTERS WARGAMING CENTER

The wargaming ops center amphitheater was crowded, its eighty-four seats arrayed in tiers already swarming with people. Twelve rows, each containing seven player positions, were arranged in a quarter-arc, all facing five large-screen displays and a long row of tables near the front. Each row's façade was capped with ample ledge space for participants' documents. All 84 stations were equipped with recessed, dual-screen computer displays, a keyboard and infrared mouse, plus a personal wireless headset for communicating with others in the room.

An oversized briefer's platform, complete with multiple controls arrayed in a cockpit-like fashion, faced the participants. A retractable LCD screen framed the briefer's platform. Aft and to one side of the podium sat an information technology technician. The podium resembled a stand-up fighter pilot's station with an engineer's jump seat added. Wargaming technicians referred to the briefer as the "Pit Boss," and Androsin was always the Pit Boss when outside experts participated in STRATCOM's wargames. The wargaming ops center was his domain.

"May I have your attention, please," Androsin bellowed, more a statement than a question. The room quickly quieted as seats filled.

"Although we've all been wrestling with the satellite outages in our various groups, we are now establishing a formal wargame with much broader objectives," the colonel opened. *And with far more serious consequences,* he thought, but didn't say.

Jill Bock was distributing wireless personal display assistants, which contained wargame objectives, scenarios, team assignments, and a schedule. Most important, the teams' "deliverables," the expected products of the group's efforts, were prominently delineated.

"Team assignments are in your PDAs, and I'll review them before we get started," Androsin continued. "You'll note that we're taking advantage of findings and insights from the DEADSAT wargame we ran here about a year ago," he explained as the room's occupants scrolled through their PDAs. "Frankly, we've been a bit surprised at how close the DEADSAT scenario parallels the satellite failures, both commercial and government, we've experienced recently. But that's actually turned out to be a positive for us. We're going to launch the current wargame from the DEADSAT foundation, in essence continuing that game, but now with near-real-time, real-world inputs. Consequently, this will be a very dynamic exercise, more than we'd like it to be, I'm sure."

Androsin surveyed the participants, ensuring their attention. "I don't have to tell you, I know, but General Aster asked me to underscore one point: What we accomplish here is critical to the nation's security. Folks, we absolutely, positively *have* to figure out what the hell is happening to our spacecraft, and who's doing it. This wargame will be like no other most of you have been involved with. Our deliverables will not remain fixed as the game unfolds, because too much is happening on a very short timeline. And our boss needs a list of options from us, specific as well as general courses of action he can recommend to the SecDef and the president. He needs them immediately, if not sooner."

Continuing with the orientation, Androsin turned to the wargame's mechanics, outlining how "blue" or friendly teams would interface with "red" teams that emulated adversaries. "Let's review our gaming process. There will be two Blue teams. Blue A will represent conventional U.S. decision-making processes and adherence to the chain of command, where combatant commanders report to the SecDef. Blue A will also have an interagency component for parallel planning and NSC reporting requirements. Blue B will represent a futures body. Its role will be to look across combatant commanders' unified, geographic and functional responsibilities, while assessing real-world intel concurrent with the Red team's moves against Blue. Think of Blue B as an integrated private- and public-sector team with an economic, diplomatic, and military emphasis. This team must think beyond our traditional roles of collecting and analyzing information, then developing recommendations for the president. Blue B will examine a range of potential near-term scenarios and, with BOYD's weighted assessments, will create a variable set of actions that can be fed into Blue A's decision and action process."

Here it was 2010, Androsin was telling himself as he segued to the next section of his briefing, and he was still explaining how another Blue team would be creating an outside-the-box strategy. Interagency processes were still bound by stovepipe thinking, where ideas and decisions went straight up the line, yet constrained by departments, divisions, bureaus, and command desks. A paucity of information-sharing across bureaucratic boundaries still predominated as zealous staffers protected their bosses, particularly when bosses were power-hungry political appointees. This failure to share information, to think collectively instead of defensively, had led to what they were now facing: linkage blindness—an inability to see the relationships among elements. *Not today,* Androsin told himself grimly. *Not on my watch.*

He surveyed the senior group's faces, knowing his wargaming expertise and organizational skills more than made up for his relatively subordinate position as a colonel. "The Green team will assume its traditional role as allies and coalition partners. And I would like to reinforce General Aster's thanks to the distinguished officers and diplomats from Australia, Canada, and the UK for representing such a critical part of our relationship." He nodded to the allied representatives in one corner of the center. All were assigned to STRATCOM as military liaisons for their nations.

"Colonel, do you have provisions here for a separate space team?" asked Jack Molinero, the TransAmSat Vice President.

"Yessir. We're going to depart from what many of you previously experienced. I've set up two space cells comprising government and private-sector individuals. You may notice in your handbook that Blue A and Green will have space subject-matter experts embedded, but we want *separate* space cells for Red and Blue B. Red will fully explore our space vulnerabilities, while Blue B explores unique ways to resolve our current dilemma—and try to stay ahead of Red."

Androsin continued, "The Red team should convene immediately. Admiral Stanton Lee will lead. But right now he is in a meeting with General Aster. Begin your problem set and we'll brief the admiral when he arrives."

Jill wondered how Admiral Lee might react to the developing space situation. He would already have information about the satellite outages, of course. The guy had sources everywhere. A retired four-star admiral, Lee had concluded a stellar career as the combatant commander of Pacific Command. PACOM, the largest U.S. geographical command, ranked equally with the Middle East as arenas chronically poised at the brink of incalculable war. Lee had distinguished himself by handling several small skirmishes between North Korea and Japan, then China and Taiwan. He out-thought and out-fought the North Koreans and the Chinese so effectively, and contained the conflict so quickly, that both countries quickly backed down after getting their noses bloodied.

In 2008, outgoing President Bush had recognized Lee's skills by naming him U.S. ambassador to the People's Republic of China, where he served until retiring in 2009. Lee spoke seven Chinese dialects, and knew China's leaders better than most. His military and diplomatic experience had proven invaluable in building a strong new coalition in the Western Pacific, the Indian Ocean littorals, and Australia. Similarly, he had dealt firmly with fundamentalist Islamist explosions across Indonesia, the Philippine Islands, Malaysia, and Thailand, rolling up al Qaeda sleeper cells as they readied freight containers filled with explosives bound for the Port of Los Angeles. Lee was a proven, consummate strategist. Having the admiral lead this critical wargame's Red team was never in doubt.

"Colonel Androsin, how are you structuring the wargame moves?" came a question from the rear.

Androsin thought for a moment, sensing he should back up a bit and explain wargaming from a more elementary level. The group included some first-time gamers, and everyone needed to be on the same page.

"Your team leaders will help you with the finer points once we begin. But so we're all together on this, let's cover wargaming principles and how we'll apply them. Apologies to you old hands. Er, *experienced* hands." He smiled. For the newbies, he oriented them to a way of thinking, reminding each participant how important challenged-thinking could be in helping to find a solution.

"Wargaming is derived from an ancient art of gaining understanding about the *intent* of one's opponent, not just his capabilities. Many of you are familiar with U.S. intel products. We still count things—numbers of bombers, submarines, missiles, divisions. You know the routine. But what about *intent?* That's much more elusive. We caught glimpses of Soviet intent during the Cold War, thanks to intercepted conversations between Politburo officials, but we missed the bigger picture. We were reactive. The other side developed a threat capability, and we countered—the classic arms race. But what about the sociology, even the anthropology, of an adversary's intent? Wargaming helps us gain greater clarity here," Androsin explained.

"But you're talking about predictions. And nobody—*nothing*—can do that," one of the commercial representatives challenged.

"Not predictions, but an exploration," Androsin clarified. "An examination of a range of possible futures. An investigation of multiple geopolitical scenarios and military actions, followed by an assessment of those scenarios having the greatest consequences, and a self-assessment of our own abilities to counter, preempt, or even survive any of those consequences."

We're literally in the multiverse, Androsin thought, *where anything is possible, just by thinking about it. Decisions can be made and unmade with the click of an electronic stylus, and entire scenarios can be loaded.* This was a stage and his teams were the actors. Only this time, the moves would be real. And people could die.

Another challenge arose. "I maintain you *are* talking about predictions, and that's dangerous territory for—"

"We are not here to debate the merits of wargaming, but to understand better what we are facing, and what we should do about it!" From the rear of the amphitheater, Admiral Lee's commanding interruption grabbed everyone's attention. His signature formal, prep-school style snuffed further discussion. Even his grammar, ringing with authority, had a crispness to it that commanded silence.

"In the absence of a wargaming discipline, most operational and strategic decisions are hammered out by default. By that, I mean, we look through the only prism we know: our experience, some intelligence or other type of information, our own judgment, or simple blinders-constrained prejudice. That can lead to considerable groupthink, especially the higher one gets in the food chain. At its fundamental roots, wargaming is about testing our judgments in a challenged forum, based on a rigorous methodology of move and countermove, developing a series of options, and retesting those options for the intended and unintended consequences. Then we recommend options that are structured as decisions for our national leadership."

As Lee spoke, Androsin was reminded of the actor John Houseman playing Professor Kingsfield, standing at his lectern, dressing down a first-year law school class in the movie *The Paper Chase*. The style was Lee's own, however, tempered during a fellowship at Princeton's Woodrow Wilson School long ago.

Lee descended the stairs, his mere presence demanding the wargamers' focus. "I recommend we get on with the task at hand. As for you doubters, cease this uninformed debate on whether we should be wargaming or not, and either put your expertise to work or just get out of my damned hair. Carry on with your brief, Colonel." The room was silent, a sea of stunned expressions.

Androsin struggled to suppress a grin. "And good morning to you, as well, Admiral Lee," he said, nodding to the now seated retired admiral. Turning, the colonel jabbed a button on his control panel, then aimed a laser pointer at a large screen filled with bulleted text. "Here's a summary of our situation as of this morning, and the questions we'll be addressing. One: Are there logical, engineering-based reasons for satellite failures? Or is something unusual occurring? We need to sort this out ASAP." He paused, glancing over his shoulder to read the admiral's face, before continuing.

"Two: *Aviation Week* magazine picked up on the rash of comsat and commercial imaging satellite failures, and published articles earlier this week about how they're affecting the industry. Besides the obvious service disruptions—the type that Mr. Molinero outlined some time ago and that are causing such a ruckus among our citizens—*AvWeek* claims that insurance rates are starting to climb dramatically. This could be a red herring, though. So far, adjusters tell us they're not worried. Evidently, they consider these satellite losses just a part of the industry's painful maturation process, regardless of what *AvWeek* says.

"Third: The CIA has relayed an unconfirmed report about several U.S. and UK military satellites being illuminated by a relatively low-power laser emanating from near Chelyabinsk in the Ural Mountains of Russia. The SecDef's office has asked the State Department to protest, and the British prime minister's people have already lodged a strong complaint with Moscow. Far as we can tell, though, there's been no damage or impact on operations of those military birds."

Androsin turned back to the podium and adjusted a thin microphone at the corner of his mouth. "Okay, let's get going," he concluded. "Rob, how about spooling us up on the Excalibur situation."

"Sure. Our latest Excalibur satellite had been up only three weeks, and was working perfectly," recounted Rob Joaquin, a Pantera senior engineer who had been with the 130-satellite, global telecommunications project since its inception. "It was fine initially, but at the next automatic housekeeping check, its power supply was showing signs of overheating, and the battery pack was slowly discharging. Solar arrays were still working, but the batteries weren't charging. We hadn't seen this before, and the system configuration was exactly the same as on fifteen other late-model Excaliburs we'd put up. All our engineering reviews and test data scans came up negative. There's no on-board reason for this particular anomaly," Joaquin summarized.

"Anything else?" Androsin prompted, a bit too sharp, perhaps. Civilians tended to get lost in the techie details. It was his job to keep the wargame at a higher, more strategic level.

"The failed bird seems to have experienced a sharp electromagnetic pulse of some kind during that two-hour period, but we don't know why," Joaquin continued, unflustered. "Not a huge EMP, though. Nothing of the magnitude you'd expect from, say, a high-altitude nuclear detonation, but still powerful." He had already been through this, in one form or another, with executives of Pantera and Excalibur Systems, the operational firm. Any of those execs could fire him in a heartbeat. This damned Army officer could push all he wanted, but Joaquin didn't answer to him.

Rob tapped a keyboard at his position, drawing the room's attention to a smaller screen behind and to one side of Androsin.

"Here's a detailed analysis and a visualization of those dead or limping satellites' positions over the Earth. We've found an interesting correlation here. Most of our failed Excalibur birds were in view from that same green area we saw the other day—southwestern Asia," Joaquin concluded. "We think somebody's screwing with our spacecraft, too."

Androsin glanced around the room, pausing as he momentarily locked eyes with the National Security Council representative, Preston Abbott. "So, exactly what are the impacts of all these so-called anomalies?" Abbott asked, his tone laced with sarcasm. Androsin had disliked the pompous little toad from day one. *Why didn't the NSC send a more appropriate representative?* Earlier, while briefing Androsin about his expectations from this wargame, Aster had confided that Abbott's presence led one to question whether the White House's chief national security advisor really understood what they were facing here.

Those damned Washington . . . , Androsin thought, a flash of red-hot anger flooding his being. *The idiots still don't get it! These space-based resources are absolutely critical to America's security, and their loss is serious business. We're not ready to tackle real-world issues precisely like this wargame is addressing. But now we're smack-dab in the middle of one. And,* the colonel grumbled silently, *unless we get the right people from the NSC, the FBI, CIA, National Security Agency, State Department and other agencies involved, we'll never get policies in place to handle space problems—or even get the current ones solved! Now we're having to figure this shit out in real time, when it's actually hit the fan. Damn!*

Abbott was in his mid-thirties, an ex-congressional staffer who had been swept into the presidential administration when his former principal had been tapped as the new commander in chief's national security advisor. He was an all-too typical political science graduate who had risen a little too quickly in the Washington scene, Androsin thought. Many lightweights like Abbott had arrived on the new president's coattails, and the damage their naive idealism had managed to do in one year was being felt by military units across the globe. Full of self-importance and drunk with perceived political power, they were short on critical practical experience, the kind of expertise and understanding acquired through mentored seasoning, especially when it came to complex technical matters. This new crop of idealistic functionaries was poised to cause yet another round of havoc. Until now, Abbott had remained silent throughout the initial troubleshooting discussions, clearly lost amid the intricacies of commercial and military space.

"Anybody like to bring Mr. Abbott up to speed?" the colonel asked gently. Abbott missed the patronizing tone.

TransAmSat's Molinero didn't hesitate. An astute judge of people, he, too, had sized up the NSC rep as being a lightweight. "Well, as far as TransAmSat is concerned, it's had mixed effects. We're providing service to our customers, but we've had to scramble. We're drawing on some backup systems, and we're trying to buy some transponder time on other people's satellites to make up for any shortfall, if one of our limping birds bites the dirt. But, as more of these die, the whole industry's hunkering down, worried that they might be next. However, everybody's hoarding their excess capacity, at least in the comsat arena."

The big man heaved a sigh and continued. "Unfortunately, that's the good news. On the business side—and by God, this had *better* not leave the room—our insurers are getting a trifle antsy, just as *AvWeek* said. They've already raised our premiums on two spacecraft scheduled for launch next month, and have reduced the risks they've routinely covered in the past. So, that leaves us hanging out more than we used to. The insurers' main gripe is that we haven't found an airtight reason for our on-orbit problems. There doesn't seem to be any common thread, and Boeing, who makes the basic HS-601 and HS-601HP busses we use, is baffled, too. If there's nothing to fix, the insurance sharks are worried that something might happen to our new birds after launch. So, up go the rates, down goes the coverage, see?" Molinero leaned forward, looking down the curved row toward a middle-aged, nondescript woman. "Have I mapped that out about right, Mitch?"

Audrey "Mitch" Mitchell, a Boeing technical representative based in Omaha, agreed, but had nothing to add. Pantera's Joaquin did, though. "We haven't seen any adverse impact on future Excalibur sales or operations yet, but we're still hunting for a fix that'll make the insurance guys feel better. So far, however, rates for Excaliburs haven't changed."

A slightly overweight but attractive Asian woman with streaks of gray in her short, swept-back hair spoke up. Androsin hadn't noticed her before, and was surprised at his lack of "situational awareness." As most of the wargame's players and company reps were men, he at least should have noticed the few women in attendance.

"I'm Barb Leewon, from EarthView," she said. "I can tell you that the loss of our EagleEye 1 had a severe impact on our company. That was our first sub-meter imaging spacecraft, and it was a major, major hit to EarthView. Even now, years later, we're still recovering from it. We laid off about thirty percent of our workforce a few months after the failure. Our CEO stepped down under investor pressure, and several other executives resigned. J. D. Hart was a little harsh in his assessment the other day, but, in essence, he was right. We don't know why that receiver nuked itself—as he so indelicately put it—and that mystery has had strong repercussions. Yes, we've tried to put an upbeat public face on the situation, but it's still serious."

Leewon's candor clearly surprised her commercial colleagues, Androsin noticed. Most of the nonmilitary participants in last year's DEADSATS wargame had been reluctant to share any of their companies' shortcomings or failures, even flatly refusing to talk about some aspects of their satellite operations. Nowadays, competition was a stronger force than national security classification. The recent on-orbit losses obviously had changed commercial-world dynamics.

"But one thing we hadn't expected was the immediate, very aggressive response from the investment markets," she added. "Within days of our declaring EagleEye a total loss, our financing literally dried up. Worse, money for the whole commercial imaging and remote-sensing industry took flight. Investment bankers simply refused to talk to us, and some of the other start-ups, as well. Money's been slowly coming back in the last couple of years, but, for a while there, it was touch and go. Nobody was confident we'd get the backing necessary to get even a single sub-meter imaging system in orbit and be competitive with the first-generation outfits, like Space Imaging. So, the loss of only three imaging spacecraft—one of NASA's Landsats, the Israeli Ofeq-9, and ours—was enough to scare off the investment houses. If direct-broadcast TV and the Big-LEO high-bandwith comsat systems hadn't come back so strong after 2006, we still might be hurting for outside funding."

Androsin was taken aback. Not exactly a business whiz anyway, he had never considered how fragile the commercial space market might be. From his military officer perspective, it seemed quite robust, since new satellites were being launched at a faster pace now than ever before, and the Pentagon had always found a taker when it needed more communications or imaging support. Hell, the commercial guys had launched more spacecraft in the last few years than the military and government-civil communities had, combined. Air Force Space Command's experts had spent months planning how to leverage the new crop of comsats and imaging birds to fill the military's ever-increasing shortfall in space-based communications, direct broadcast and imaging requirements. The seemingly endless global war on terrorism, in particular, was stretching military resources well beyond what the Cold War had demanded.

In a flash of insight, the Army colonel became acutely conscious of exactly how tenuous the U.S. commercial space sector really was—and the national security implications of that fragility.

8 APRIL/NATIONAL CITY, CALIFORNIA

More than fifteen hundred miles away from Omaha, Zipporah Moffitz was thinking about her parents in Haifa as she tuned out the monotonous droning of a self-important FBI analyst at a chalkboard in the front of her conference room. Zipporah was a Sabra. Her great-grandparents had walked, ridden mules, and begged rides on merchants' wagons as they made their way from their tiny *shtetl* in Russia after one of the czar's pogroms, across Turkey, and into the Middle East. Through successive generations of rabbis, scholars, at least one archeologist, and farmers, Zipporah's family weathered the changing winds in Palestine. One of her uncles fought in the British Arab Legion during World War II. Another uncle had served with Montgomery at El Alamein. And yet another had flown with one of the Hashemites on a Lancaster bomber in the Royal Air Force over the skies of Dresden and Hamburg.

And in her own generation, coming of age in the 1940s when the British had shut down Jewish immigration into Palestine to satisfy the Arabs, Zipporah's older brothers had fought with the Hagganah and the Palmach. It was reputed that one of her cousins had collaborated with the Stern Gang when they set an explosive device in the King David Hotel. And Zipporah herself, after graduating from Stanford, had returned to work as an analyst for Israeli banks and, occasionally, for Menachem Begin's Likud party. All while she kept her academic credentials intact, teaching at the University of Tel Aviv one year and at New York University the next. But Zipporah led a double life.

She was Mossad.

Her FBI file, still highly classified, read that field agents had tracked her with "foreign intelligence operatives" through the Manhattan subway system, photographing various entrances and exits, "paying particular attention to Times Square." Her file, heavily redacted for all but agents at the supervisory level, also mentioned that she had worked as an analyst for both Israeli intelligence and the NSA. She had been temporarily assigned for a year to the Department of Justice for white-collar crime analysis, not only at the Hoover Building in Washington, but also at the FBI Academy in Quantico, Virginia.

And she had worked as a consultant, even while teaching in Israel, for the DEA, helping set up systems to track drug cartel proceeds through the international banking system and back to Colombia. On more than one occasion, she had identified Cayman Island banks that served as safe havens for drug cartel money managers as they commingled cocaine dollars with clean dollars before routing funds back into the United States. She was good at what she did.

The conference room was unbearably hot, but this meeting would soon be over, Zipporah kept telling herself as she inched closer to the window air-conditioning unit at the National City offices of the DEA. She hated task-force work. Mostly talk and all interagency policy, each agency representative covering his butt, so that male egos up the line wouldn't get bent out of shape. Tread lightly, she had been told. Don't argue with the Feebs and don't step on military toes. The Navy CID types from San Diego who had been brought into DEA and Customs especially didn't like reporting to female academics, especially the uppity ones who acted like they belonged in the field, when in actuality, they were rarely out of the classroom. She couldn't let on that she worked for Mossad and had slept with a Sig Sauer on many a night.

New York City cops—now *they* were different. She spoke their language, albeit with an Israeli accent. She had lived just south of Washington Square and sometimes walked downtown for early meetings with NYPD intel managers who assembled in the One-Five squad room to scrub wiretap data on the Gambinos, the Genoveses, the Westies, and whatever other gang was selling weapons on the streets of lower Manhattan. In fact, she loved New York City detectives, especially the cops in the DA's special investigative unit. If she had to go anywhere ugly, she would rather have a New York street cop covering her back than any federal officer. These were *real* cops.

The caffeine was wearing off now, and the feds' interagency policy protocols had begun to echo inside her brain like the whirring of the grinding head of a dentist's drill. *Unbelievable what doesn't get done*, she grumped to herself. Her coffee was stale, the Styrofoam having imparted its own distinctive flavor to the swill, and the person in charge of making a new pot had somehow managed to disappear. She hated the doughnuts, was desperate for a no-kidding breakfast, and was now officially claustrophobic.

Then, over the pounding of the air conditioner's compressor, she heard the lilt of a thick outer-borough accent that made her so homesick for her adopted city that she wanted to cry.

"The lawr is clea-uh," the voice said slowly and emphatically, trying to erase as much of its New York dialect as possible, but not succeeding. The speaker was dressed in a soggy half-sleeved white shirt, tie pulled down under an open collar and hanging off to one side, loosely hanging khaki chinos, and brown wingtips. The man would have been remarkably unimpressive had it not been for the menacing .44 tucked into a shoulder holster under his armpit. Not federal issue, for sure. She could almost smell that this guy was also carrying a private piece in an ankle holster.

"Been that way since the eighties. Banks gotta file cash transaction reports for every transaction over ten long."

A few eyebrows went up over the obvious mob cant. But who cares, she thought. *These Midwestern moral-majority types need a little street exposure.*

"So your attack on the case begins with the institutions where there's no cash transaction report, even though you know there's gotta be one."

This was anti–money laundering 101, the basic introduction to how cash flows from the street drug-cell leader through the local money manager and into the banking system. And the NYPD detective—formerly one of the department's top homicide investigators, a lieutenant from Manhattan North, then an intel officer, then the chief of the cold-case squad, before retiring to work on joint Customs Department and DEA money-laundering task forces—was laying out the basics. But he wasn't talking about the drug cartels today, which was why Moffitz was there.

Today, Detective Lieutenant Frank Donovan, retired, was talking about a private banking system used by Middle Eastern families for hundreds of years. It financed all kinds of businesses for cousins, uncles, worthless sons-in-law, and nephews across continents and oceans. It was how families increased their wealth. And in an age of computer tracking and digitized bank accounts, it was a way Hamas leaders, from the Gaza Strip and West Bank, to jihadists in Afghanistan had managed to move huge sums of money right beneath the noses of the counterterrorism units around the world.

"Ya see," Donovan explained. "They don't use computers, at least not in the way we do. Like the old dons in the crime families back in the day, they keep the numbers in their heads, write them down on slips of paper with no other notes, but never on anything we find that makes sense to us."

For all the interbank tracking and the investigations of Iran-Contra back in the 1980s, the real terrorism money was passing through private family banking systems, where an Uncle Khalid or Mustafa might be the loan officer, making his tallies and payments on a sheet of paper only he could retrieve, and making sure that a distant cousin fulfilled his obligation to the rest of the family. But now the banking system was routing money to distant cousins leading cells in jihads against the West and against the Zionists. And Zipporah Moffitz was *the* expert on tracking the money, even money running through legitimate Western businesses.

Ever since the UN Security Council had caved in to the Chinese in 2006 and allowed the Iranians a token right of uranium enrichment, thus saving face, while enabling the Russians to handle the commercial enrichment program on their behalf, Mossad had gone to wartime intel footing. Its operatives had infiltrated every aspect of Iran's covert enrichment programs, but still hadn't figured out how Iran planned to keep some operations—those that had to be done above ground—hidden from the all-seeing eyes of Western spy satellites. Then, in 2008, it all became suddenly clear. Too bad they couldn't tell anybody about it. And that was why Moffitz was at today's meeting.

When Western satellites began going dark, mysteriously disappearing into radio silence, someone deep inside Israeli intelligence—it wasn't Moffitz—put enough twos and twos together to ask the question: Who benefits? His question had an easy answer: Follow the money. Wasn't that what Felt, AKA "Deep Throat," had told Woodward and Bernstein almost forty years earlier? Follow the money. Who benefits?

Clear enough to figure out who benefits from imaging-satellite losses. Anyone on the ground who didn't want something to be seen would benefit, *if* the surveillance satellites went dead. But the money? That was a different story. How can you get something to pay for itself? And that was where Moffitz came in. She had to go where satellites were born to ask the question.

Donovan was wrapping up. His charts had followed a hypothetical dollar through the round—from the street to a coffee importing business, to an offshore investment bank, where dollars were sold for euros, and then back to Colombia as profits from a Hong Kong electronics company. But today, he said, they were tracking Iranian money. How was money getting into the U.S. in such quantities that even Customs and the IRS couldn't track it?

"Somethin's wacky," he said. "We can't figure out the source of the money. So we look for somethin' else happenin' at the same time. Maybe there's a tie-in, even a remote tie-in." He pointed at Moffitz. "Zippie, yerrup."

Gotta love New York City cops. She'd never met Donovan, but somewhere he'd read her file or spoken to another cop at a downtown bar. She had been nicknamed Zippie by the cops at the One-Five.

"We have a situation in the Middle East right now," Moffitz explained, speaking very slowly so that even the feds would understand what she was saying, given her Israeli accent. "We know the Iranians need to move their military assets above ground. But as soon as they do, our satellites will see them. So how can they hide them? Only one way. These satellites have to go blind. And that's what's been happening for almost three years. But that also takes secret cash, black money that the banks don't see. We know where every drop of their oil money goes, but we don't know about private family banks and where all *that* money goes. Is it coming back into the States? How? Where? Through what kind of front business? Can we correlate that to the aerospace business?"

A Customs agent popped up. "And we're tracking a huge counterfeit operation out of the Middle East right now," he said. "Huge. In dollars—tens, twenties, and fifties—enough to go through a 7-Eleven store without too much trouble. Tracked it in cities where there's large hack and limo businesses, where drivers can make change quickly and people won't do a double-take. Talking Vegas, New York, Dallas, Houston. Airport cab traffic. Some of it's coming in from South America, too."

Moffitz was thinking on her feet now, juiced by the corollary activities. Satellites going silent, money coming into the U.S., a large counterfeiting operation out of the Middle East. Somehow this stuff would tie together if they could only find the conduit.

The Customs agent sat down and Moffitz began to think aloud, as if she were lecturing at the FBI Academy. She asked if there was a tie-in between the satellites and the influx of money, and if the influx of money had to do with Iran's converting to euros instead of dollars, and if someone was putting counterfeit dollars out there, while another country was leaving dollars big time . . . how did it all fit?

This got Donovan up, pacing rapidly. "Sure it fits," he said. All they had to do was find the source of an influx. But more than that, they had to tie the source of the dollars, both real and counterfeit, to some entity in the U.S. that also was tied to satellites. "Look fuh the connection," he said. "It's in plain sight, right in front of our noses. You just have to follow the money. Who makes money in the satellite business?"

The question hung in the room like the aroma of bubbling soup in the hallway of a Lower East Side tenement apartment building.

Follow the money, Zipporah repeated to herself.

4

OUT OF THE BLACK

Hassan Rafjani allowed the faintest hint of a smile to curl the corners of his thin, tight lips. He stood near the sterile white wall of a laboratory in the heart of his country's Isfahan nuclear fuel–processing plant, watching Iran's popular and charismatic president deliver yet another venom-laced speech that was calculated to shock the West's decadent leaders once again. President Mahmoud Ahmadinejad's words appeared spontaneous, as did his wild, almost erratic, gestures. But Rafjani had carefully written each word and choreographed each movement, then coached the nation's latest figurehead to deliver them with well-staged fire and vehemence.

Rafjani's incongruous, piercing blue eyes strayed, quickly taking in his surroundings. The sophisticated facility was the best-known nuclear site in Iran, but by no means the most vital. He, Rafjani, had masterminded the strategy and charade that encompassed what now played out before him. Virtually singlehandedly, he had presented Isfahan to the global community as *the* icon of Iran's growing nuclear might, while the real work—the development, subcritical testing, and production of Iran's handful of nuclear weapons—was quietly carried out at Parchin, an extremely secret, remote underground location in the Persian desert. Although important, Isfahan was largely for display, a titillating peep show to keep nosy international Atomic Energy Agency inspectors occupied. And it had served that purpose well for more than five years, refining nuclear fuel needed for the handful of weapons recently moved to Parchin, yet also serving as a focal point for the American, European, and Israeli bastards who would deny Iran and Islam their rightful share of world power. In essence, Isfahan was the magician's gloved hand and waving kerchief that held inspectors' attention as the other performed the actual magic of weapons development.

A chorus of enthusiastic shouts from the small crowd of nuclear scientists, engineers, technicians, and white-gowned production workers drew Rafjani's attention back to the speaker. The president was in excellent form today, ranting like a madman, flecks of white spittle at the corners of his mouth, as he exhorted his nuclear minions to "raise Persia to the destiny God intended! You are chosen ones, the soldiers of God who will rain His fire down on the American heathens and Zionist occupiers who enslave Muslims! You are the dreams and hopes of all Iranian people. . . ." The president's red-faced, angry litany continued, the tone rising at just the right time and rate, the raised fist pummeling the air above his head in exclamation. He was good, very good, and the television cameras were capturing it all.

He, Rafjani, had chosen well, then nurtured, coached, and prodded this charismatic, malleable, ambitious man, ultimately creating one of the most feared national leaders since Hitler. And the Iranian president would soon be the undisputed spokesman and spiritual leader for several billion Muslims across the globe, all committed to wiping every godless, materialistic infidel off the Earth's sacred surface.

But the true genius behind this president was Hassan Rafjani, and he was virtually invisible outside Iran's tiny cadre of power. Although the American CIA, Israeli Mossad, and dozens of other Western intelligence agencies knew *of* him, none had any inkling that Rafjani was the strategic architect and political power that charted Iran's course. None had recognized that *he* was the shadowy mullah who told the president and the well-known, more visible mullahs what to say and do. His brilliant mind held absolutely no doubt that he, Hassan Rafjani, was chosen by God to fulfill Iran's destiny. He knew that his life and work had been predicted by the prophets. He was the ominous "man in the blue turban" mentioned in the cryptic Nostradamus quatrain, the mysterious Mideast leader who would trigger a world-cleansing firestorm, a decades-long war, a precursor to the return of the Twelfth Imam.

The CIA had given him a code name that Rafjani cherished, a moniker that, even now, warmed him: "Dagger." Very fitting, he thought, because it had been his own razor-sharp, wavy-edged, jeweled-handle stiletto that had loosened a captured CIA agent's tongue so long ago. The agent had bled profusely before finally cracking and, in ragged, halting whispers, dumped memory about Rafjani's Dagger file: tall, dark-skinned, angular, but with unusual pale-blue eyes that hinted of possible European ancestry. Deep roots in Iran's radical Islamist community. One of the students who had kidnapped more than sixty Americans during the Shah's overthrow in 1979, holding them hostage for 444 days. The most cruel and feared of all those captors. One of the then-Ayatollah's most devout, loyal servants.

But what the intelligence agencies failed to recognize was that, over subsequent decades, the CIA's deadly Dagger had steadily risen to a position of considerable power within the tight circle of clerics who controlled Iran's political and military cabal. Also unknown to those supposedly superior intelligence services, Dagger was the force behind America's recent satellite losses, the architect of the money-laundering and counterfeiting programs, and the genius who would soon unleash the Great Satan's ultimate and complete destruction.

Listening to Iran's president conclude his speech, his voice rising to a well-rehearsed crescendo, Rafjani held a single thought: *The time of Iran's destiny has arrived.*

9 APRIL/"GATOR" AIRSTRIP/FLORIDA PANHANDLE

With a finger in each ear, Brigadier General Hank "Speed" Griffin waited on the hot tarmac for the Air Force C-21 Learjet's TFE731 engines to shut down. Enough of his hearing had been lost flying and hanging around fighters most of his career, and the flight docs had warned him that he'd better get serious about wearing ear protection on noisy flight lines, or else.

The twin turbofans whined to a moan, a drop-down passenger door opened, and Army Lieutenant General Dave Forester, STRATCOM's operations director, emerged, an aide wearing Navy khakis in trail. Griffin met the three-star general with an outstretched hand. "Speed Griffin, sir. Welcome to nowhere-land."

Forester grinned and nodded. "Dave Forester. I think we crossed paths at the Pentagon, didn't we?"

"Yes, sir. I was an air staff puke, and I believe you were on the Joint Staff then. Thanks for coming down and taking a look at this system, sir. It's, well, different, but it may be just what General Aster needs right now," Griffin said, leading the senior officer to a car parked near a nondescript building. Griffin wore a USAF-issue olive-drab flight suit adorned with only a leather nametag and a single white star embroidered on each shoulder. The usual command and unit patches were pointedly absent, Forester noticed. *Typical spook outfit,* he thought.

"General Aster didn't say much about your system, just that we need it for immediate space support. He sent me down to get a first-hand look, since my guys'll probably be the ones tasking it. So, what do we have here, Speed?"

"Sir, it'll make more sense when you see it—and we're headed there now—but, in short, it's a two-stage-to-orbit spaceplane system," Griffin said as both men ducked into the car's backseat, using opposite doors. The air-conditioned interior was a welcome relief from Florida's humidity. Although the April day would be considered mild by Southern standards, the outside air was still uncomfortably hot and steamy, a result of the surrounding swampland pumping a stifling, smelly moisture into the air, nonstop.

"You said two-stage? You launch an expendable booster from *here?*" Forester's sunglass-shaded eyes swept the airfield, which was surrounded by tall pine trees and thick undergrowth.

"No, sir. The first stage is just a giant airplane that gets us out to Mach 2 or so at high altitude, then we drop the spaceplane. The little bird lights its advanced-technology engines and climbs into space, while 'mama' flies back home, usually here."

"Exactly where the hell *is* 'here,' Speed? My C-21 pilots weren't at all convinced there was even a runway out in this swamp."

Griffin laughed. "Sir, they're not the first guys who've said that. We're still on the Eglin AFB reservation, but this strip isn't on any flight charts, at least none you'd find in a flight ops chart file. This base has only one road that runs through miles of swampland, and the place is surrounded by a tall fence topped by razor wire. The 'gators out there are outstanding guards, too. We have one runway—a damned long one, for sure—but our camo-and-deception guys did a great job of disappearing it. Unless you know where to look, you'd never spot it from the air. That's why we tell first-time pilots to contact our controllers on a very specific, protected frequency . . ."

"Which doesn't exist, either, according to my pilots," Forester added.

"Right, an unpublished frequency. Once they check in, though, we basically talk them down until they can finally pick out the runway visually."

"Is that because we're not trusting GPS these days, at least for precision landings?" Forester asked. "The GPS ball is sick and getting worse, you know. I suppose General Aster briefed you on the situation?"

"Yes, sir. He did. And, on my day job, I've been tracking both the satellite losses and their impacts on national security operations for the last few days. Of course, I've been down here for about, oh, forty-some hours," Griffin said, checking a large chronometer on his wrist. "Anything new along that line, sir?"

Forester stared out the window on his side for a few seconds, then turned back to Griffin. "Speed, the GPS ball is continuing to degrade. We're trying to mitigate the timing and position inaccuracies for high-priority military users by jiggering with their receivers' software, so that bad spacecraft are ignored. Each satellite has an identifier, so we can tell high-end receivers to ignore those particular inop birds. Of course, we can't do that for every receiver out there. Only the DOD ones that we have immediate access to.

"And the rest of the system?" Forester continued, punctuated with a deep exhale. "Frankly, it stinks. Several commercial comsats and imaging spacecraft are down. We've lost at least one DSP, an old missile-warning platform, and we think a few of our sigint birds have been damaged. We're hurting."

"Any new leads on what's causing this garbage, sir? Again, I've been out of the loop. . . ."

"Aster's got some experts sequestered in a heavy-duty wargame at headquarters, and they're definitely making headway. Those folks are absolutely convinced we're dealing with no-shit attacks, too. Their whiz kids have ruled out technical causes. They've narrowed down the area where attacks are coming from, and we're looking hard in that zone. We sent an Air Force U-2 mission over the area, and we've had a Global Hawk UAV parked nearby for a solid twenty-four hours, but nothing's turned up, so far." Forester reached for the door handle as the car stopped beside a massive structure. A small knot of civilians and Forester's aide waited, huddled in a narrow pool of shade thrown by the building's overhanging roof.

Griffin introduced his guest to the civilians, then opened a small walk-through door. Several very tall, now closed, sliding doors to his right rested on a set of parallel tracks nestled in the concrete apron. Forester stepped over a six-inch raised threshold and ducked to avoid the door's steel overhead structure. Straightening, the three-star stopped and stared.

"Hoooly . . . that's the mama ship?" He was staring upward at one of the biggest, most sleek aircraft he'd ever seen. A long, pointed nose arced toward him, cantilevered well in front of a long, multitired nose-gear strut. Highly tapered wings swept back from the graceful fuselage, their tips flowing up and back, forming tall, graceful winglets. The entire aircraft spoke of paint-blistering speed. Cavernous rectangular engine inlets tucked under the central fuselage would gulp prodigious quantities of air, feeding four powerful turbojets buried far to the rear. The entire vehicle gleamed, painted a glossy white.

"Yes, sir. When it's carrying the spaceplane, we call the whole system 'Blackstar,' but the official designation is the SR-3/XOV. You're looking at the SR-3 carrier aircraft, or mothership," Griffin explained, leading Forester beneath the forward fuselage. The vehicle's belly was still well above the men's heads. "This puppy is about 200 feet long, and was built from long-lead structural items left over from the old XB-70 supersonic bomber program in the sixties. After one airplane crashed, the program was killed, although NASA flew the other ship awhile for research purposes. But a lot of structural elements for a third aircraft were already built, so they went into storage."

"There's one of these in a museum somewhere. The Smithsonian, right?" Forester asked, craning his neck to peer into gaping main-gear wheelwells embedded in the delta-shaped wing.

"There's an XB-70, the only one in existence, in the National Museum of the Air Force at Wright-Patterson AFB in Ohio, sir. But SR-3 is an advanced, modified version of the XB-70. It can carry a conformal, external load under the belly over there," Griffin said, pointing to a concave depression in the fuselage, between the main gear. "See these fairings? They were added to control airflow when we're carrying the XOV out to high Mach numbers.

"Over here," Griffin continued, steering the small group toward the forward fuselage, "we have retractable canards, one on each side. These provide lift during slow-speed flight, but they also compensate for a fairly substantial center-of-gravity shift when the spaceplane is dropped at high speed."

"Impressive," Forester said, obviously awed. "Now, why was this system developed, and why's it still so highly classified?"

"I wasn't around in the early days, so I'll ask Earl here to fill you in on that. I came on board as one of the spaceplane test pilots in the mid-nineties," Griffin added, turning to a slight, stoop-shouldered man wearing a white, open-collared golf shirt.

"Well, general, back in 'eighty-six, right after the Challenger shuttle disaster, the Air Force had a bunch of rocket failures," Earl explained in a thin voice that bespoke his eighty-plus years. "For a while there, this country couldn't get a damned thing into space. Really had the big boys worried. All of a sudden, DOD didn't have what they call 'assured access to space' anymore. Those boys were fit to be tied, I tell you. 'Bout that time, I was playing golf with an old friend over at Rockwell, part of Boeing now. I was a designer for the Skunk Works, Lockheed's advanced projects outfit, you know. You bein' an Army guy and all, you might not be familiar with this aerospace stuff."

Forester flushed and glanced at Griffin. Generals weren't accustomed to being patronized so blatantly, even by elder-statesmen contractors. Speed caught the senior general's eye and shook his head imperceptibly, silently indicating: *Let it go. I'll explain later.*

Earl continued, unfazed, his hands thrust into baggy slacks as he rocked fore and aft on the balls, then heels, of both feet, usually staring at the floor. "Like I said, we were playing golf, but really, we were just trying to pry stuff out of the other guy about what was going on at each other's company. That's what those golf games were always about, ya know. Anyways, we got to talkin' about all the rocket and shuttle problems—Rockwell built the shuttle, ya see—and what could be done about 'em. Well, that led to kickin' some ideas around, so we gave up on golf and settled down to a beer over at the club.

"Now, you get a couple of old engineers together over a beer, with plenty of napkins layin' around, and things just start happenin', ya see? Next thing ya know, we had ourselves the makin's of a two-stage-to-orbit spaceplane system. The Air Force took some convincing, but those boys were desperate, so we eventually wound up with a crash program. Long and short of it is, we built two of these SR-3s—one of 'em crashed later—and a couple of spaceplanes, one unmanned and another manned one. Rest is history, general."

"I see," Forester said. "But what convinced the fly guys that they needed this, the Blackstar system?"

"Coupla things. For one, the blue-boys needed some way to get stuff into space lickety-split, something they and they alone could grab and use. They'd been led down the primrose path by NASA and the politicians, ya see. 'The space shuttle will always get your payloads into space, so you don't need those damned expendable rockets anymore,' NASA told 'em. Well, the Air Force didn't quite buy that, so they kept some Titans and Atlases around, just in case. But in early 'eighty-six, some of those started blowin' up. Blue-boys were in a hell of a fix. So, we convinced some government big shots that we could give 'em a reliable, although limited, way to get some things into space on short order. And that was Blackstar.

"Two, they liked the system's flexibility," the aging designer continued. "See, we could get this system off the ground pretty damned fast, then shoot the XOV into any orbital inclination we wanted. That way, we had complete surprise the first time around. You might not know this, but satellites are awful damned predictable. People on the ground can figure out when one of those critters is gonna fly over, so they run out and throw a tarp over whatever we'd really like to get a look at. Well, by God, our system took care of that! On the first go-round—the XOV's first orbit—we could catch 'em with their pants down, see? Bad guys had no way of knowing the XOV was headed their way. We had a good electro-optical package that slipped into the spaceplane's Q-bay, and it was somethin'. Take pictures of the hair on a turtle crawlin' across the highway. Damn sure did the trick." Earl paused and scratched his chin, glancing at Forester to see if the general was still tuned in. He was.

"And third, we could get close to other guys' satellites, long as they were in low orbit. Air Force thought that was pretty damned slick. Never know when you might want to slide up and take a close look, right? So there ya have it."

Earl's abrupt summary left the small group silent, waiting for the next epistle. But Earl was finished. Nodding curtly to Forester, he turned and wandered away, apparently lost in thought as he surveyed the SR-3's underside.

"That's it?" Forester asked, incredulous.

Griffin grinned and shook his head. "Earl's a character, sir. But he knows more about this system than any living soul on Earth. So, we tracked him down—he's retired, living down near Miami—and convinced him to help us reactivate it."

"Aster said it's been mothballed for, what? About ten years?" Forester asked. The two generals were headed for the huge hangar's opposite wall, trailed by three of the civilians and the Navy aide, who carried Forester's briefcase. Earl was left by himself, squatting under the SR-3 and peering up into the depression that normally held the XOV spaceplane.

"Yes, sir; about ten years. I left the program in ninety-nine, when it looked like the Clinton Administration was going to pull the plug on it. We'd been having a few technical problems before then, and some White House bean counter decided he'd spend the money elsewhere. I won't bore you with the details, but suffice to say the whole system didn't live up to expectations, and it was damned expensive to keep all this infrastructure running for the relatively few flights we were able to conduct. That last year I was here, I only flew four times, and the other pilot had about the same number of hops."

Forester's brow furrowed, confused. "So, why crank this up again now, if it didn't work?"

"Oh, it worked, sir. Worked damned well, in fact. But not up to what some of the program's salesmen had promised. Because everything was *so* highly classified, it only took a couple of chiefs to decide, 'This isn't working out. Kill it.'

Especially when Washington's hatchet-guys made it clear Blackstar was a goner. I don't know your politics, sir, but my take on it was: Here we had a president who didn't care much for the military, in general, and who wanted to spend a lot of money on social programs. Social programs were going to be his legacy, the next Franklin D. Roosevelt, I guess. I doubt if he ever knew about the SR-3 and XOV, but one of his hatchet guys did. And axe-man's charge was to kill everything he could and put the money where the prez wanted it. Killing this program was easy. No constituencies to scream bloody murder. No jobs lost on Main Street.

"The kicker, though, was that we'd had a major breakthrough on the technical side, and this system was starting to pay off big time. We ran some real-world ops, and they were *very* successful. I'm not at liberty to talk about them, but you probably saw their results. Still, the system's earlier, not-so-great track record was reason enough to kill Blackstar. We were able to get bean counters to okay mothballing the vehicles and retaining a skeleton force to keep the lights on, but that was about all."

Griffin opened another walk-through door and flipped on a light switch. It took a few seconds for overhead lights to sputter to life, slowly dissipating the room's darkness. The effect, though, would have made a Hollywood director proud. The half-light slowly brightened, revealing a low-profile, sleek craft worthy of a science fiction movie. Forester whistled softly. He might be an Army helmet-head, but he could still appreciate cutting-edge technology. And this animal was definitely cutting-edge.

"This is the XOV-2, sir. The manned version," Griffin announced, circling the vehicle. A small ladder-stand on the opposite side led to an open cockpit, its relatively small, thick-windowed canopy raised to a 45-degree angle. "Want to climb in?"

Forester carefully worked his six-foot three-inch frame into the tight cockpit, and found himself staring at several large flat-panel screens on the instrument panel. All were dark now; the vehicle was unpowered.

"*This* is abso-frickin-lutely amazing, Speed! It really goes into space?"

"Damn sure does, sir. I've flown it up there several times. Quite an experience. The contractors went ahead and upgraded some of these avionics after the shutdown, so this Dash-2-*plus* model is even more capable than when I flew it last."

For the next twenty minutes, Griffin, aided at times by the three Lockheed Martin and Boeing civilians, described the XOV-2's systems and capabilities. Forester climbed out of the cockpit and walked slowly around the Star Wars–looking spaceplane, clearly awed by the machine. Finally, the two men stood apart from the civilians, admiring the craft from its nine-o'clock position.

"Okay, Speed. I'm convinced this thing is real, and I'll brief the boss on what you have here. But tell me, exactly what can it do to help us resolve the mess we're in now?"

"Sir, Blackstar was designed for three primary missions: quick-reaction, surprise recon-surveillance, which is what Earl described, emergency reconstitution of space-based assets, and covert attack. We've proven it can do all three. I guarantee it can be a huge asset right now, primarily by helping us restore some critical space capabilities. We've rounded up a few micro- and nanosats—small, experimental tactical satellites that perform very specific functions—and we can put them into precise low-Earth orbits. A few of those quick-response payloads are being readied for flight at government labs as we speak, and I expect them to be on site here in a matter of days," Griffin explained.

"And the attack capabilities?"

"That's a little tougher, sir. Getting the weapons, at least of the type we used before, is problematic. We're working with the contractor, and he's busting his tail, but we won't see them for a couple of weeks, at best."

"What can we do with weapons? From space?" Forester, as Stratcom operations chief or "J3," was familiar with all of the nation's long-range strike forces, both land-based and submarine-launched missiles, and the B-52, B-1B, and B-2 bomber fleet. All had their limitations, politically as well as operationally. Griffin explained the in-space-deployed weapons and their value, sometimes in hushed tones to be sure the nearby civilians wouldn't overhear. They might be cleared for the technology, but not the classified tactics of weapons employment.

Griffin concluded his brief explanation and grinned. "Tungsten rods, sir. These things come smokin' out of the sky, do their thing, and there's not a sliver of anything left with 'U.S. of A' stamped on 'em. Vaporized. Totally covert. The bad guys never know what hit 'em."

Forester nodded slowly, again looking at the spaceplane with unabashed awe. "Speed, we damned sure need this system, and soon. What're your limiting factors and schedule? How soon can we put it to work?"

"Sir, the 'limfac' right now is fuel for the XOV. It's an endothermic gel, and really tricky to make. There's only one contractor who can do it, and he's scrambling to reconstitute an in-house capability. All his equipment was kept in working order over the last ten years, but the smarts—the people who really understood the black magic required to make the stuff—they're scattered all over. The company's trying to pull a couple of the key chemical engineers out of retirement and round up some experienced technicians, then get 'em all out to northern Utah, where they made the fuel."

Griffin hesitated a moment, then added, "This fuel is boron-based, about the consistency of toothpaste, and manufacturing it requires a very precise, tricky process. Pumping it where it's supposed to be in the XOV without freezing solid or turning into gunk is *really* tough, especially throughout a flight's temperature profile. I'm hoping that the company can start making a healthy supply of it soon, but getting the right people back . . . That's the one thing I'm sweating blood over, sir. We had no money to keep the smart folks together, so they went elsewhere. We've literally lost the intellectual capital needed to make this fuel miracle happen. This whole *system* is a no-shit technomiracle, sir, but the fuel tops everything else."

"That's the American way, Speed. Make the scientific and technical breakthroughs, then piss away the intellectual capital that made them possible. All because of the almighty dollar!" Forester expounded forcefully. Griffin was surprised at the Army officer's fervor, but agreed.

"Speed, I want you to do whatever it takes to get this system ready to go, and soon. I'll get you the bucks. SecDef is blasting all the black-world hurdles that've kept it under wraps, and he expects results. He's extremely upset with the spooks, so I hope fallout from his rampages doesn't interfere with what you're trying to do here. If it does, let me know. General Aster is indebted to you for bringing Blackstar to his attention *and* for kicking those majors in the butt so they'd spill the beans to the Boss. In fact—you'll appreciate this—I heard the secretary say during a teleconference: 'When were you hyper-patriot bastards going to bring your super-secret systems out of Area 51? After the nation was on its back and down for the count?' He's using up a lot of political silver bullets for us. So, I can't emphasize enough, we are in dire straits right now, and we *will* damned sure need Blackstar," Forester stressed.

"Understand, sir. Until a couple of days ago, I had access to the president's daily intel briefing—part of my real job at NSC—so I had a sense of the world situation. Any recent developments I should know about? Something that might help us motivate folks to get Blackstar operational?"

Forester nodded. "Afraid so. There's some intel—still sketchy, but compelling, just the same—that seems to indicate Iran has picked up on our space vulnerabilities, and their mullahs are getting frisky. We're trying to keep an eye on them, but . . . Well, you know damned well how some of these space-platform losses have screwed up the normal acquisition and flow of information." The general extended both arms, bending the fingers of both hands inward. "If you think of our nation's ability to monitor geopolitical activities, what China, North Korea, Iran, and terrorist organizations are up to, it's like a big jigsaw puzzle with most of the pieces in place. That gives us a pretty good picture of what's going on around the world. But when we lost a good-sized chunk of our space-based assets, it was like removing dozens of pieces from that puzzle. We no longer have a good picture of the world situation. Our situational awareness has been drastically degraded, and we're having to *guess* what's going on in Iran, China, and elsewhere. Not good," Forester said, folding his arms again.

Griffin nodded, listening intently. He did *not* like what he was hearing.

"We need to reconstitute that capability, and *fast!* The U.S. definitely needs much better SA than we have right now," Forester concluded, then snapped his fingers. "Oh, there's another important tidbit that came in just before I left Offutt; doubt if you've heard about it. A National Security Agency sigint bird intercepted a cell-phone call that the spooks think might be related to our space problems. Some Iranian bad ass they call Dagger was discussing some kind of 'big strike' with a drug cartel chief in Colombia. 'Space' was mentioned, but this Dagger guy jumped all over the druggie's butt, reaming him about saying too much about the 'big op' on a phone. That caught one of the NSA analyst's attention."

"Any idea what they were talking about?" Griffin asked.

"Not really. Our wargamers' best guess is there's another hit on a satellite coming, but no telling when or what system might be targeted. Whatever it is, the Colombian said the hit would be 'spectacular'—and there's not a damned thing we can do to protect the target," Forester added, shaking his head in frustration. Griffin only nodded in agreement.

"We also think China's got something going—not sure what—and there's some increased chatter in terrorist circles, too. Looks like every damned bad actor out there wants to take advantage of our space problems! My boss is convinced we're approaching a perfect storm on the geopolitical scene, where almost any untoward event could blow up and become a serious issue—diplomatically, economically, and militarily. Who knows what's coming next? The SecDef and president are worried about the same thing. This shit is serious, Speed, and getting more so every hour, I'm afraid."

The Army three-star pointed at the low-slung XOV resting on a special cradle designed for long-term storage, preventing a nose gear and outrigger skids from contacting the concrete floor. "By the way, do we have crews for these birds? Who's going to fly them when you *are* ready?" Forester asked, shifting the subject.

"Sir, I'll have to fly this one, at least initially," Griffin said, unable to suppress a grin of anticipation.

"You gotta be . . . you mean to tell me a *general* is the only pilot in the whole damned Air Force who can fly this bird? That's a little hard to buy, Speed!" Forester's bullshit flag was clearly elevated and waving.

"Only two pilots were ever checked out on the XOV-2, sir. It didn't fly much, so we only needed two guys for flight testing and the few real-world missions we flew," Griffin answered cryptically. "The other pilot left the service after the program was mothballed. Fortunately, he's now working for the *right* contractor as a test pilot, and the company's agreed to get him back over to Blackstar as soon as they can. He's flying for a black program out in Nevada at the moment, and it's also under a very high-priority time crunch."

"Higher than SecDef demanding we get Blackstar in the air—that is, into *space*—ASAP? Come on . . ." The BS flag was still waving.

"Sir, I don't know what the hell he's flying, and I'm not sure 'flying' is even the right word, as advanced as that thing might be. But the SecDef made it clear that 'Zulu,' the other XOV pilot, will *not* be available for several weeks. That leaves me as the only body to fly the first few missions."

"Hmph. Okay. If you say so. How about the other bird, the SR-3? Got a crew for it?" Forester cocked a thumb toward the huge hangar bay they'd been in earlier.

Griffin nodded, a lopsided grin appearing. "It's ready to go. One of the really clever things the NRO did at the outset of this program was to *not* buy the Blackstar system. They sniveled some CIA 'black' money—who knows where they got the other few billion—then paid Lockheed and what was then Rockwell to build two SR-3s and two XOVs. But the deal was structured so the companies retained ownership of all vehicles. There was never a DD-250 signed, transferring the birds to the government. The companies provided all-contractor crews for the carrier ship, and the unmanned

XOV-1 was flown remotely by contractors, as well. There were only two military pilots involved. We both flew the Dash-2, the manned spaceplane."

"A turnkey-contract operation, huh?" Forester completed the thought, nodding. The three-star stared at the steel-gray spaceplane.

"Exactly," Griffin continued. "Once the system's existence leaked out, the press started asking hard questions. Most of the mainstream press called it 'Aurora.' 'Av Leak' didn't use the 'Aurora' moniker, but they were especially hot on Blackstar's trail in the mid-nineties. They badgered everybody from SecDef to the NRO director to Space Command, at the time. But because it was a contract operation, everybody in the government chain could honestly say 'we don't have *anything* like that!' Nobody had pictures or an inside source, I guess, so . . . voilà! Blackstar didn't exist! Nobody lied, either. The federal government really did *not* own and operate a two-stage-to-orbit spaceplane system. Drove the newsies nuts! They never did figure it out."

Minutes later, Griffin and the civilian cadre shook hands with Forester before the STRATCOM operations director climbed aboard his C-21. The three-star was back in Omaha a few hours later, briefing Aster about Blackstar and giving the STRATCOM chief renewed hope.

And yet, both generals admitted to a gut-level feeling that the nation was on the threshold of unprecedented calamity. Unfortunately, neither recognized that events on the other side of the globe were already spinning out of control. And there was nothing they could do to prevent the nightmare about to erupt.

9 APRIL/STRATCOM WARGAMING OPERATIONS CENTER

Jill looked at Androsin with admiration, maybe even a little affection. Then a new data point on the screen caught her attention.

"Colonel! What's this about a new capability?" she inquired.

"Hang on, Jill. I gotta get these teams moving!" Androsin was trying to collect a pile of documents from his desk in the gaming lab.

"Jim!" Jill insisted, knowing the use of his first name in a relatively public arena would grab his attention.

"What is it?" he asked, exasperation coloring the response.

"What's this new capability? BOYD says we have a regeneration device of some kind."

"Come on Jill, we ain't got time to play with future space systems!" He was hurrying, trying to get back to the gaming center.

Jill looked at her computer again, then turned the screen to show Androsin, a knowing look of near-mischief highlighting her good looks.

The colonel impatiently grabbed her laptop, scanned it for mere seconds, then ran to his workstation to answer a clanging red phone. He dropped several papers en route.

Jill retrieved the sheets, quickly perused them, then nonchalantly proclaimed, "Sooo, we really *do* have a spaceplane in play. *There's* our regeneration capability!"

Aster was on the other end of that red phone. Androsin nodded, "Yessir, we're convinced that none of our usual suspects are behind the DEADSATS situation. A couple of wargame participants are now suggesting terrorist or criminal activity; maybe both."

"Look for some way to identify the source of these satellite troubles," Aster ordered. "Follow up on that NSA cellphone intercept. Throw that into the mix. Make sure our commercial folks are thinking like bad guys, *any* kinda bad guys. Hell, bring in some real bad guys if you have to! Get those people to stretch their brains to capitalize on all the BOYD-generated probabilities."

"Roger, that," responded Androsin. "Admiral Lee already has the Red Space cell working a non-state angle to all this."

"Good. And prep some potential responses. I want an immediate and effective option to execute once we've pinned down who's behind this DEADSATS operation."

Androsin hung up, returned to Jill's station, and spoke in a low, even tone. "Load the latest intel, then get BOYD to talking. We need analyses of these latest GPS anomalies ASAP. I've gotta get back to the gaming center. I'm afraid we're falling behind the power curve," he concluded, heading for the door.

In the center, Stanton Lee was addressing the DEADSATS wargamers. "There will soon be several Red actions in play. The current action against Blue's—ah, America's—satellites is most likely attributed to a criminal and/or terrorist cabal."

He paused for effect, nodding toward Androsin as the colonel entered. "Red factions, which represent communist China, North Korea, Iran, and even Russia, are busy conducting their own assessments of Blue's situation. As Red, we are preparing options to take advantage of the rapidly spreading vulnerabilities faced by Blue."

Androsin sighed and muttered, "Oh shit." If the wargame was any indication, the global space situation was rapidly deteriorating.

5

VULNERABILITY

*The era of procrastination, of half-measures, of soothing
and baffling expedience, of delays, is coming to a close. In
its place, we are entering a period of consequences.*
        Winston Churchill (Pre–World War II speech)

9 APRIL/REMOTE IRANIAN DESERT SITE

Ari, you sure nailed this one, the undercover agent breathed.
Half-buried under a greasewood scrub bush, he slowly
swept the well-camouflaged desert outpost with a high-
powered night-vision monoscope. The incessant wind had di-
minished somewhat, but small clouds of sand occasionally
whipped across the few hundred meters separating his hidey-
hole and the feverish activity of about thirty uniformed soldiers
and technicians.

Ari, the agent's sometimes flaky Iranian operative, had relayed a tip that "something big" was happening here, a tip the
agent had almost blown off. Ari's transparent attempts to
wheedle a few more euros—it always *had* to be euros, not
dollars—from his handler had bred well-earned skepticism.
But this time, something in Ari's frightened, darting eyes had
lent a twitch of credibility to his claim. Consequently, the
agent had invested several days in making an apparently

meandering trip to this godforsaken patch of Iranian no man's land. Tonight, Ari's tip was paying off, big time.

The middle-aged agent snapped a few digital infrared photos to augment his legendary memory for details. *These'll convince the skeptics at Langley,* he mused. No question about it: Those troops were preparing a Shahab-4 missile for launch, and doing it in such a way that an intelligence-gathering satellite would never spot it. From here, he could look directly into a camouflaged softly illuminated tunnel and see the long missile resting on a transporter-erector-launcher vehicle, ready to roll. But the heartstopper was its payload. He was 90 percent sure that those technicians were mounting one of Iran's newly produced, but precious few, nuclear weapons, developed right under the noses of all-too-complacent International Atomic Energy Agency watchdogs. Something about the almost reverent, super-careful way they handled the "package," monitored closely by several high-ranking officers, was a dead giveaway. No way to tell for sure, but he'd bet his pension he was eyeballing a nuke.

What the hell are they getting ready for? Whatever it was, it scared the pee-wad out of him. He had to get this info to the right people, and fast.

Satisfied that he'd gleaned all he could, the agent carefully slithered backward from his hiding spot, rolled onto his back and methodically, slowly made a horizon-to-horizon scan with the night-vision scope. Nothing moved but wisps of wind-driven sand. Still, he waited, relying on years of danger-honed senses. He trusted the unexplainable feelings, the hunches, as much or more than he did his five physical senses. Time and again, those hunches had saved his butt. Tonight, though, the hunches and feelings were quiet.

Moving slowly, he walked a long kilometer back to his concealed car, conscious of staying on hard-packed desert soil as much as possible. Yes, the wind could be depended on to erase footprints. Unless you positively had to depend on it doing so. That's when the fickle wind-god would leave a perfect, telltale print in exactly the spot where a roaming security patrol would find it, and maybe him, as well.

A few hours later, with the eastern sky hinting at a new day, the CIA agent, known only as "COBI" to all but a few at Langley, shut the car's engine off. He was still a hundred meters or so from his nondescript rented room, but chose to walk the distance. No need to wake the owner and raise unwanted questions. The bearded, dark-skinned agent entered, then silently closed and locked his room's door. He pulled a battered-looking case from a shelf and placed it on the bed's lumpy, sagging mattress. Removing a small, very thin laptop computer, he rapidly typed a stark, definitive report, appended the digital infrared pictures he'd snapped earlier, then saved everything as a coded file ready for transmission.

He double-checked that a covert, ultrawideband transceiver set up near the window, yet shielded from someone that might pass by outside, was talking to the laptop through a high-speed cable. The omnidirectional antenna, buried in the battered case's outer, removable shell, was tilted to the proper angle, aimed to hit an invisible satellite several hundred miles above. The agent waited until the laptop showed a faint READY icon, indicating a weak but usable link had been established with at least one Excalibur spacecraft. He tapped a series of keys and hit ENTER. A brief burst of covert pulses shot skyward, seeking what he hoped was a still-functioning space platform that would dutifully relay his critical information to CIA headquarters halfway around the world.

Aw shit! Shit! he muttered, unconsciously holding his breath. An expected confirmation indicator failed to appear. Only an icon that guaranteed the message had been transmitted continued to blink. But had it been received? Two more tries elicited more curses, but no confirmation. Then the satellite was gone, out of reach. Just one Excalibur had been in view at this particular hour. Normally, there would have been several of the birds available, all cross-linked to guarantee robust communications for any ground-based user. The agent quickly repacked his high-tech gear, double-checking that the case's ingenious, X-ray–proof compartment was secured. For several long minutes, he sat in the darkness, thinking.

This shit's way too important to rely on hope, as in "hope it got through." Risky or not, he had to get out of the country, get to a secure, fiber-optic land line. He'd been alerted just days ago that he could no longer depend on the Excalibur satellite links. Because the commercial Big-LEO constellation had lost several of its spacecraft in recent weeks, coverage at certain times was spotty, at best. Unfortunately, it was the only satcom system compatible with his state-of-the-art equipment, gear given to only a few, highly trusted, deep-cover agents.

Why the hell don't we have a backup system? he grumbled silently, now lying on the bed, staring at the ceiling. He knew the answer, of course. Renting transponders on commercial satellites was cheaper than building a new generation of high-cost, government-operated National Technical Means space platforms. Since all covert agents' messages were cleverly encoded to look like innocuous e-mail traffic, why not go cheap? Besides, the private-sector's satellite technology was more advanced than what the feds could afford and field anytime soon—even if the new administration in Washington *was* inclined to put up the bucks. And it wasn't, he knew. But tonight, when he, one of the CIA's most valuable, proven agents, absolutely *had* to get a critical-to-national-security message out, that cheap, high-tech link had failed.

Unable to sleep, he quietly packed the few things he had left lying about earlier in the evening to assure any snoopers that he'd be back, and splashed cold water on his face. He scanned the room one last time, ensuring nothing was left to betray who had been there, then exited. His waiting car's engine still temperature-ticked as it cooled. Turning the ignition key, he was relieved that the finicky vehicle started once again. He drove west, toward the border, wondering: *If Iran is really getting ready to lob a nuke at Israel, Europe, or who knows where, and I can't get a warning to my people . . . Damn! That "cheap," pay-as-you-go comsat service could wind up being very, very expensive. And incredibly deadly.*

9 APRIL/CIA HEADQUARTERS: COMMUNICATIONS CENTER

The shift supervisor scanned a decoded message he'd been handed and scowled. "What in the hell is COBI trying to tell us? Looks like we got only part of his transmission."

"Beats me. That's why I brought it over," the comm technician said, shrugging. He stole a glance at the super's watch. The day shift was about over, and he was ready to move out. "Uh . . . want me to have second-shift run it through again?"

"Won't help to run it again. 'Digital don't lie,' remember? This is as good as it'll ever get. I can't make sense of it, but it's still tagged top-priority. I'll run it up to ops." There were a lot of weird messages flying around these days, and the operations director wanted to see anything from that part of the world, and pronto.

Neither the CIA supervisor, nor the agent COBI in Iran, now driving through the early-morning dawn toward the Iraq border, could have dreamed what that partial, scrambled message would unleash.

9 APRIL/STRATCOM HEADQUARTERS/COMMANDER'S OFFICE

"General Griffin on line one, sir."

"Thanks, Annie," Aster said as he swept through his assistant's outer office. He grabbed the walkaround transceiver from his desk, then tossed a flight jacket on an overstuffed chair.

"Speed? Thanks for getting back so quickly, bud. Listen, we're in *more* trouble. Buzz's people just received unconfirmed reports of attempted sabotage at two of our overseas satellite ground stations. Buzz is worried about our domestic sites being hit, as well."

"Whoa, you're outrunning me, sir. Guess I should know, but who's 'Buzz'?" Brigadier General "Speed" Griffin held a cell phone to one ear, trying to juggle a stack of documents in one hand as he left the White House by its west door.

Aster slid into the leather chair behind a large, glass-topped desk, then spun to face an expansive window. Offutt AFB's grass was finally starting to green up, he noticed absently. "'Buzz' Sawyer. Air Force Space Command chief. Came out of missiles, then worked space most of his career. Surely you've run into Buzz. Probably when you were up there at air staff."

Cursing under his breath, Griffin wrestled a red-covered, TOP SECRET–emblazoned document back into the stack under one arm. "Nope, never met the guy. Hell, Howard, guys like you and me . . . we never crossed paths with the pocket-rocket troops! Too bad, too. We didn't learn about this space business, like they did. Just fly the jets, get our tickets punched, do a couple of wars, get the medals . . . shit! Who'd have guessed we'd have to get space-smart? Fighters were all that mattered!"

"What the hell are you blabbering about, Speed?" Aster mouthed a thank-you to Annie, who had placed a steaming cup of coffee within reach.

"Space, dammit! Our so-called aerospace team— airplanes, satellites, spaceplanes. All one nice, seamless continuum from five hundred feet to orbit. Well, we don't have a seamless 'aerospace team' now, and we're damned sure paying the piper for it!"

Griffin was definitely wound up today, and a little irritable, Aster observed with a grin. The normally unflappable fighter-turned-test pilot was obviously back in Washington, which probably explained the grouchiness.

"Yeah, okay. So we should've let the Marines have space, but it's too late for that," Aster chuckled. "Right now, we're in a fix, and the bad guys are lining up to take advantage."

"Get anything out of your gamers yet?" Griffin asked, finally dropping the pile of documents on his desk. He kicked the small office's door shut.

"Some preliminary stuff," Aster said. "Lee's Red team suspects some kind of criminal or terrorist action is behind our satellite problems, or a combination thereof. He's downplaying the odds of a nation-state being the culprit. BOYD's orbital analysis points to some possibilities in the trans-Caucasus region. But there's no better correlation than that right now, except for that NSA cell-phone intercept."

Griffin squeezed into a creaky government-issue gray chair wedged between a battered bookcase and his equally dull gray desk. One-star generals detailed to the National Security Council staff rated nothing better in the Old Executive Office Building pecking order. "General, I just left an update briefing on the space situation," he replied. "The president wanted to know how bad things really are, and was asking tough questions about our systems' status, both commercial and national security. And, by the way: he was *not* pleased with J. D. Hart, that NASA loose cannon out there in your wargame. Evidently, Hart was blowing off steam on CNN yesterday afternoon. C'mon, Howard! A 'Pearl Harbor in space'?"

"Hey, Speed. If I could control every civilian player in this game who has access to a newsie, my title would be 'STRAT Dictator.' Truth is, old Hart probably did us a favor by getting some straight information out there. And we damned sure don't have time to be staffing responses for the sake of political correctness. It's time for action, not politics!"

"Well, you're gonna get action, but you're not gonna like the politics that come with it, sir. The president just approved two of your wargamers' options, and said he wants the context of Lee's assessments fed into a nationally televised address his speechwriters are drafting right now. If they can get air time, you can expect to see him on the tube tonight."

"*Our* options? And those options are . . . ?" The volume of Aster's voice jumped substantially.

"That UN Security Council resolution option. The one calling for a halt of all support to those nation-states responsible for malicious interference with space systems."

"Damn it, Speed! We discarded that one! It's *not* a viable option! It was as toothless as a month-old baby! That Security Council resolution idea was nothing but an initial BOYD dump. Once our teams went through their paces, it was discarded. That thing was never supposed to go forward to Washington!" Aster was pacing now, growing more agitated by the second. Even the *domestic* political situation was spinning out of control!

"I figured that was the case, Howard," Griffin continued, trying to calm his ex–flight leader. "But some NSC staffer got his mitts on *all* the options being kicked around that DEADSATS wargame, and they wound up here, in the wrong hands. By the time I found out about 'em, it was too late to cross-check with your guys."

"The friggin' UN will take too long, and won't do diddly, anyway!" Aster fumed. "Look, Speed. Whoever has our spacecraft in his sights damn sure has the Kentucky-windage down pat. And he'll keep shootin' till we're on our knees. Those spacebirds are like ducks sitting on a pond. We *have* to go with a space *démarche!* And if that doesn't resolve this problem, we'll have to start shutting down Russian and Chinese space assets with our nonlethals, just for our own protection."

"There's more, Howard, and I'm sure it's more to your liking: The prez has green-lighted special ops missions to take down the source that's killing our sats—just as soon as you locate it, of course. STRAT will have operational control of those SOF teams. You'll get some prep assistance from European and Central commands, but it's your mission."

"Good! We damn sure expected to run this one. How's State faring with the Russians?" Aster stared out the window toward the flat Nebraska horizon, but without seeing it.

"They've got a tentative okay for any necessary counterspace ops, but the Russian interior minister, Gospov, wants to send observers," Griffin said, then moved the cell phone away from his ear, knowing what was coming.

"What?" Aster practically screamed. "Oh, for christsakes! That is *not* gonna fly! I don't want *any* damned Russians out here, looking over our shoulders when we're running counterspace ops. *Understand?"*

"No way, sir. That won't fly with this president. He's looking for a way out of this mess, and is already working political damage control. Having the Ruskies involved is his way of playing the international-cooperation card. You should hear the bullshit flowing around these hallways!"

"Good grief! Look, Speed, I have no control over how the political poop goes down in Washington's fantasyland. All I can tell you is that we all know what happened back in 2003, when the Russians had their own intel units in Iraq slipping Saddam info on our tank deployments. Even the Russian ambassador was spying for Hussein. And the White House wants to bring Ivan into this loop? My job is to keep this country from being blindsided by a surprise 'Red' attack, and that's a much bigger near-term issue than Washington's blame game. My gut tells me this whole situation could go south if somebody like China, Iran, North Korea, al Qaeda— who knows?—feels lucky and decides to jump into the fray while we're limping."

Aster noticed Annie standing near his elbow. She held up a scrawled note for him: "Gaming Center. Now!"

"Gotta run, Speed. Do whatever it takes to yank any mention of response options from the president's address tonight. We're not ready to send messages via the do-diddly UN yet, and we damn sure don't want to tip our hand to whoever's behind these attacks!" Aster was steaming, incensed that political expediency might trump his and others' efforts to contain the imminent crisis before it exploded into *many* crises.

"Rog, sir. I'll be in touch," Griffin signed off. He flipped his communicator closed, and for a few moments, let a wave of fatigue wash over him. Flying a one-of-a-kind, aerospike-powered rocketplane into orbit and back safely was a no-sweat undertaking compared to what Aster had just asked him to do. Convincing White House political animals to pull those UN options from a presidential speech, especially at the eleventh hour, would require all the charm and persuasive powers he could muster. *Too bad they didn't send us fighter pilots to charm school. But this is why they pay me the big bucks . . . I guess. . . .*

In Omaha, Aster jammed the secure-line handset into its cradle. Hands on hips, he quickly scanned the commander's battle screen on one wall of his large office, trying to divine a big-picture understanding of the complexity of unfolding events. But electronic imagery and data couldn't penetrate the veil of ominous, hidden forces he felt were covertly mobilizing against America and her allies. He *had* to find out who was responsible for attacks on the nation's space infrastructure, and why.

In parallel, he had to pull together a sense of what objectives an adversary might be pursuing. Who might want to take advantage of a situation where the U.S. was crippled, denied of its superior space-based abilities to monitor changes in military force postures? A situation where the U.S. could no longer listen to shadowy terrorist cells plot another 9/11–like attack? Where it couldn't immediately communicate with its own powerful Army, Navy, Marine Corps, and Air Force units scattered around the world? Somebody out there was committed to destabilizing day-to-day American society and the very heart of U.S. national security.

9 APRIL/NATIONAL SECURITY ADVISOR'S OFFICE

Paul Vandergrift, the president's chief national security advisor and director of the National Security Council, scanned four intelligence documents arrayed on the polished top of his oak desk. An idea was taking shape as he absorbed the dry, clipped details, mentally integrating the seemingly diverse data into a whole. Taken together, the disparate reports prompted a surge of gut-level excitement.

Don't rush. Look at it again, he chided himself, forcing calm on his now skyrocketing emotions. On the surface, his nascent idea was brilliant, a *coup de grâce* that easily could propel him into a front-runner presidential candidate slot for the 2012 election. Yet it also could make his current boss, President Pierce Boyer—*mental midget though he is,* Vandergrift snorted—look absolutely heroic today. And when the boss looked good, Paul Vandergrift's star would continue to climb on the Washington scene.

But he had to be sure he wasn't missing a subtlety in the intel reports. This was too important to be tripped up by some innocuous tidbit he'd overlooked. He grabbed a yellow legal pad, outlined a few notes as he reread all four reports, then carefully thought about each entry on the pad:

1. Satellite losses: commercial, military. Who? Why? Where from?
2. CIA field agent: Iran missile/nuke prep? Rep Guard? (Incomplete info?) Not 100%!
3. NSA intercept: telecon btwn. Iranian mullah and Colombian narco chief. Subject: space. "Big" satellite strike? Cartels involved? How?
4. Iranian president: new call for destruction of U.S. and Israel via "fire from God."

Finally, Vandergrift nodded confidently and tossed the stack of papers onto the desk. He spun his expensive leather chair and stared at the Washington skyline framed by a large window, flipping and catching a thick gel pen as he mentally reviewed his options. Iran was clearly on the threshold of a major action of some kind. Somehow, the damned Iranians were involved in all these satellite kills through some kind of cut-out or proxy, even though those STRATCOM knotheads in Omaha hadn't pinned it on Iran. Aster's wargamers kept sniffing around the 'Stans, when the evidence clearly pointed to Iran! Iran and some Colombian cartel. Good deniability. Whatever. Let Aster and his keyboard jockeys stumble about. Paul Vandergrift would take matters into his own hands, just as he'd always done.

Spinning the chair again, he snatched a crypto-protected handheld communicator—the best model Uncle Sam's spooks had developed to date—from his desktop and flipped open its cover. He punched a memorized code, waited briefly, then spoke rapidly. Moments later, he smiled, satisfied, before ending the conversation with a pithy order.

"Make absolutely sure those nose-pierced renegades of yours have perfected that worm, then turn it loose. I want

'Sting Ray' swimming through every damned computer in Iran's air defense system within . . . seventy-two hours, max. Got it?" Vandergrift listened a moment, then snapped the phone's cover shut.

He again stared out the window, thinking but not seeing. Finally, he opened the phone again and called a very private number. "Charlotte! Paul Vandergrift here. How's my favorite Canadian snowflower?" He laughed openly, knowing the chauvinist term grated on the former Canadian ambassador to the U.S. As usual, she snapped an equally biting response. They traded a few polite niceties, then got down to business. Vandergrift quickly outlined his admittedly aggressive, daring idea, and asked, "So, what do you think, dear?"

Silence dominated for long seconds, but he knew this woman. He'd wait, saying nothing, as she mulled over what he'd proposed. Charlotte Adkins, a tall, attractive, forty-something dynamo from Calgary, had successfully smoothed once-strained relations between the two North American neighbors during her 1990s ambassadorship. Smart, calculating, and a talented negotiator, she now served on the boards of several U.S. Fortune 500 companies.

Vandergrift and Adkins had met at one of Washington's many cocktail parties for the international diplomatic elite, and quickly found that they shared a common vision of what the world *could* become, given the right leadership. They each enjoyed good-natured intellectual sparring, using the other as a like-minded sounding board for sometimes radical concepts many lesser lights would find outrageous. But those stimulating jousts ultimately cemented an unspoken pact between them, a rock-solid understanding: Both were destined for greatness. And they *would* change the world, together.

"Paul, just exactly what do you hope to accomplish with this gambit?" she finally asked, her tone surprisingly soft.

He dived into the details of a sweeping, daring strategy, underscoring how critical the current geopolitical situation had become. "We're on the brink of war, Charlotte," he said. "We've been blinded in space. So we can't see that the Iranians might be playing something very deadly even as we speak. But I think I have a way to throw in a monkey wrench to buy us some time."

Then he painted a vision of what *might* be, and how she and he *might* emerge as international peacemakers. Then he sunk the hook, knowing Adkins thrived on long-shot challenges. "Of course, you'd be the Christian going into the Muslim lion's den. It could get dicey, Charlotte. If you're not up for that, just say so. Believe me, I'll understand." Again, a long silence. Vandergrift leaned back and eyeballed the ceiling, absently flipping and catching the rotating ink pen, waiting.

"One question. Why me, Paul?"

"Because you're Canadian . . . and because you're the best, Char. You know damned well I can't send an American. Canada still has a link to Iran, even if it's not a formal relationship. Ahmadinejad may be a crazy bastard, but he's still Iran's president, and he's smart enough to know he's yanking on the tiger's tail. Somehow, by playing a game too rich for his blood, he's worked himself into a no-win corner. If we give him a choice between being eaten by the tiger—and probably destroying his nation in the process—and leading billions of Muslims to a new prosperity, I think he'll choose the latter. We just need to have the right person present a fourth-quarter perfect opportunity. And you're the only one I can trust to make that offer." Again he waited, unconsciously holding his breath.

"Okay. I'll go," she said. Very simple. No arguments. No buts. Vandergrift was stunned. He had been prepared for Charlotte's endless "what ifs" and tortured, twist-and-turn analyses of pros and cons. It was her way, even though he knew she'd say yes in the end. He couldn't believe she'd caved so quickly.

"Uhhh . . . great! Terrific!" he stammered. Quickly regaining his composure, he set up a meeting for that afternoon.

"The usual place?" Adkins asked mischievously. Vandergrift smiled at the not so subtle undercurrent of her staged, husky query. He and Charlotte had never stepped over the line of propriety, knowing it could end both of their ambitions and careers if uncovered by the vicious Washington or Ottawa press corps.

Still, the sexual attraction was there, a palpable strain hovering just below the surface—and she loved to tease, knowing damned well it drove him nuts.

He laughed easily, set a time, and signed off. He glanced at a heavy grandfather clock in the corner of his spacious office, swore softly, and raced for the door.

9 APRIL/STRATCOM WARGAMING CENTER

Admiral Stanton Lee marched purposefully from his wargaming station to where Androsin and Jill Bock were loading yet another data set for BOYD's consumption.

"Colonel, I suggest we now merge the Blue and Red space cells and conduct an open review of where we are with the latest orbital track and anomaly correlation process," the admiral declared in the formal manner that had become his trademark. "The lady from EarthView . . . ahh . . ."

"Barb Leewon?" Androsin offered.

"Yes, Ms. Leewon. She is on to something that may help us refine our search for the source of interference, the actual location of the device. She is checking with the president of an outfit called Intelligent Land Management. Guess they use a process that complements existing military and national technical means. She'll brief us all when she gets the info."

"Thank you, Admiral. We'll review BOYD's data sweep for any info on that outfit ASAP." With a curt nod, the retired four-star naval officer pivoted on his heel and returned to his own station. Lee rarely, if ever, indulged in gratuitous niceties, superfluous details, or empty embellishments. *Just the facts, ma'am. Just the facts.* Androsin smiled, watching the trim, white-haired retired admiral. But his musing was quickly interrupted.

"Hey, Jim. Take a look at this: BOYD already has a read on ILM," Jill Bock said.

Seated at her own workstation, she leaned in toward Androsin as if to say, "for your ears only."

"They've been supporting some Homeland Security-

DEA–related missions, narcoterror kind of stuff," she whispered. "Not sure how their technology might apply here, but they've got the pedigree, for sure."

"We'll take any help we can get," Androsin declared. "The Washington crowd is getting mighty nervous because the politicos know they gotta do something about this space mess to calm the public down. Nobody back there's working the *real* problem, though. 'Looking good' is trumping real guidance and decision making. Hell, the entire NSC staff's tied up prepping the president for a speech tonight instead of devoting serious efforts looking at Iran, China, the 'Stans, or some other threat out there!"

Jill shrugged. "They gotta do something to calm the waters. D'you see the *Fox News* clip that came across our open-source screen about an hour ago? Seems the talking heads are picking up on J. D. Hart's 'Pearl Harbor in space' comment on CNN. That really fueled the fire!"

"Yeah, I did. But that doesn't affect us. We're not in rhythm with the damned D.C. crowd; can't afford to be. They're running open-loop, practically oblivious to the fact that we're trying to get critical info for the president, as well as some options *for* action. That disconnect is gonna cost us. Mark my word."

Jill pointedly checked her wristwatch. "Colonel, may I suggest that you get back into the pit ASAP? The admiral's working our space teams, but everybody else still needs a recap of the latest BOYD-interface protocols. We can't afford to have 'em skip over the fine points."

"True. Why don't you join the space teams as their BOYD advisor? Make sure they understand what they're getting from ol' BOYDTRIX. Try to speed things up, too, Jill. I'll stay with the Blue and Green teams, working on developing response options." Androsin exhaled deeply, fatigue flashing across his eyes as he jammed his notes into a game folder. "If we contain the situation soon, and can keep all these meltdowns confined up there in space, maybe we can prevent any shooting. I don't want anybody's mother getting a telegram."

"On my way," Jill affirmed. "Let me grab SIVTRIX first.

It'll help the teams assess overall space-system health. And Jim, if you ask me, our ground segments are really vulnerable, too. I reviewed some BOYD runs and matched them against SIVTRIX's vulnerability matrix. If someone starts picking off satellite ground stations, we're dead. We could lose control of any number of national security space assets. Not good." She scooped up a laptop computer and caught Androsin headed for the ops center.

Barb Leewon entered the amphitheater just ahead of Androsin and Bock, scanned the crowd, then joined Lee at his station. The two huddled for a few minutes before approaching the pit.

"Colonel Androsin," the admiral said, "Ms. Leewon has some important info about ILM's capabilities. To take advantage of them, though, we'll need immediate assistance from your battle staff." With a few clarifications from Leewon, Lee summarized a concept for bringing ILM into the mix immediately by contracting with them to scan the suspect land areas. "If their M-55 'U-2-ski' platform can, indeed, pinpoint the source of our DEADSATS problems, an SOF team should be in position to react quickly," he added.

The old warbird's still thinking waaay *out in front of us average bears,* Androsin thought, admiringly. Lee's suggestion made infinitely good sense. "Roger, sir. I'll alert the battle staff watch commander and have him work the ILM surveillance data into their planning for special ops team insertion." Androsin nodded, then turned to Leewon. "Barb, I really appreciate your taking the initiative on this ILM angle. Would you give our gamers a quick overview of what we can expect from ILM?"

Androsin assembled the players, and turned the floor over to Leewon. She highlighted ILM's key capabilities, outlining the company's methods for detecting environmental disturbances in the atmosphere. "Nothing penetrates the atmosphere, up or down, without leaving some sort of trace, a disturbance. A high-power beam of energy transmitted from the ground to a satellite, theoretically, would create such a disturbance," she explained.

STRATCOM's battle staff watch commander stepped into

the center, then focused on Leewon as she continued. "As Admiral Lee said a few minutes ago, this isn't about looking *at* the Earth to find the site itself. Rather, ILM can look for the environmental impact a beam-type weapon might create, such as a change in atmospherics: air chemistry, air temperature, whatever. When whoever's behind our problems fires that device, it's like poking a hole in the fabric of the atmosphere. Granted, there's little persistency, but—"

"Pardon me. You're telling us we have to wait for *another* hit before this ILM bunch can find the bad guys?" the battle staff watch commander interjected forcefully. Tempers throughout STRATCOM headquarters were getting shorter, Jill observed, glancing at the officer.

"Not necessarily. It's reasonable to expect the source will conduct periodic test firings, probably at a lower energy level, to check their system's operation," Leewon explained calmly, dismissing the commander's skepticism. "Depending on power levels, the duration of a test, when it's conducted, and dozens of other what-ifs, ILM's sensors may or may not detect an atmospheric disturbance. But yes, your concern *is* a distinct possibility," she conceded. "We may have to suffer the impact of another operational shot, and there's no guarantee ILM's sensors will even see it."

"But it's one more tool that might help us, so we're *going* to use it," Aster's firm statement boomed from the center's upper mezzanine. "Like the muzzle flash of artillery at night, it might help us find whatever's killing those birds." The STRATCOM commander had entered the wargaming amphitheater from its rear doors, then descended to the front row before turning and facing the wargamers Leewon had just addressed.

"Finding that weapon site is critical, people. This ILM capability you just mentioned, Barb, is one more arrow in our quiver. We have a lot of overhead assets scouring the area you-all identified the other day, and every tidbit of data's being scrutinized as it comes in. We're *going* to find that SOB, but we have to use every tool at our disposal! And a special forces mission is poised to take out that bad boy's site, just as soon as we pin down its location. But we're depending on you, *this group,* to help us find it. Then we'll nail it."

Turning to the battle staff watch commander, Aster directed, "When General Forester gets here, tell him I want the mission plan to integrate . . . ahh . . . the environmental outfit . . ." Aster snapped his fingers.

"ILM, sir," helped Androsin.

"Right, ILM. Do whatever it takes to get ILM moving on this. I want their Russian bird in the air ASAP, understand?" Aster demanded, searing the watch commander with a withering stare.

The four-star turned back to Androsin. "And, Jim, make sure you integrate ILM's data with all the other intel/recon stuff, as well as your BOYD data. Make sure it's provided to all the gamers here, just as soon as it's correlated. I don't want one byte of information sitting on some damned analyst's desk! Get it in here where it can be put to good use immediately! Once we sniff out that site and get a solid geolocation, we'll have the SOF guys pounce on it." He turned to Admiral Lee, pointedly drawing him into the conversation. "A team's already being pre-positioned in that general area," he said, acknowledging that this is just what Lee had suggested earlier. "Now let's get this game back on track. Time's a-wastin'." The general took a front-and-center seat and nodded for Androsin to take over.

Androsin scanned the room, noting how crowded it had become. The center now held both DEADSAT II wargame participants and several members of the STRATCOM battle staff. The air seemed to crackle with tension.

"Okay, here's an update to the wargame baseline," he began, pointing to the first bullet on a large-screen plasma display. "We've had several things happen, just in the last few hours. One: U.S. intelligence, working with their Russian counterparts, has narrowed down the region where those disruptive electromagnetic signals are coming from. The source is somewhere in the vicinity of Dushanbe, the capital of Tajikistan." He stepped aside and laser-pointed to a second plasma screen, where a map of Central Asia appeared.

"Two: our tech-spooks are working a little overtime," Androsin continued, evoking a ripple of chuckles. "Thanks to preliminary engineering data about on-orbit failures—most graciously provided by commercial companies represented in this room, I might add—the CIA's techies believe the kill mechanism is an unusual, high-powered electromagnetic beam of some kind. But they can't agree on *what* it is. Some say it could be a high-power microwave weapon, but others say no way, you can't control a microwave beam that tightly over hundreds of miles. Bottom line, they're still working on ID'ing the weapon. But that's neither here nor there as far as we're concerned." He turned and faced his audience.

"Three: CNN, Fox, MSNBC, the *New York Times*, and most of the big news outlets have picked up on that *Aviation Week* article about our comsats and other space casualties. They're all over the White House and now some newsies are camped outside *our* base's front gate here. The head-shed wants us to tell 'em something, *anything,* to calm the taxpayers down a bit. So, our public affairs folks have scheduled a press briefing at 1430 today," Androsin continued.

"'Scuse me, Jim." Aster stood, swept the crowd with a hard gaze, then pointed to another Army colonel, seated in the amphitheater's second row. "Matt, I'd like you to handle that briefing, okay?"

Colonel "Matt" Dillon, the bull-necked, shaved-bald Army Space Command chief, grimaced. The stocky all-American West Point fullback had distinguished himself seven years earlier, directing critical Army tank engagements during Operation Iraqi Freedom, the second war in Iraq. He was armor through and through, and the rough edges of a seasoned combat veteran were evident.

"You sure about that, sir? I don't have much use for newsies," the intimidating officer growled. In the midst of a heated, time-critical dash to Baghdad during a blistering Iraqi sandstorm in 2003, Dillon had literally dumped his embedded CNN reporter and cameraman on a roadside. In Matt's opinion, the reporter's stories erroneously depicted the Army tankers as aggressors and trigger-happy mavericks.

Dillon later explained, "That dumbass's so-called news clips were negatively impacting my soldiers' morale during combat operations. Therefore, I exercised my prerogative as on-scene battle commander and removed the morale-degrading element." End of subject. Dillon was roasted by inside-the-Beltway news pundits, who screamed "Foul!" But, in the halls of the Pentagon, the then-major Dillon quickly became a cult hero.

"About time you had an opportunity to talk 'space' with the world's TV audience, Matt," Aster said lightly. "Frankly, I want to send the message that we're taking these satellite losses seriously, and you sure as hell look serious, Matt!" A half-grin creased Aster's tanned features momentarily. "Get a team together, sit down with our public affairs troops, and come up with a story we can sell. Keep the particulars a little thin, but enough to send the right message."

Dillon nodded, still tight-lipped. "Exactly who are we sending this message to, sir? Besides the taxpayers, I mean."

Aster searched the room until he located Preston Abbott, the rotund NSC representative. During his brief participation in the early stages of DEADSATS II, the general—and half the room's occupants, it seemed—had taken an intense disliking to the NSC toadie. "Mr. Abbott, don't you think we should put a little subtle media pressure on the Tajiks, just to let 'em know we've pinned down the interference to their neck of the woods?"

Abbott blanched, swallowed hard, and stammered, "I . . . I don't . . . uh, let me think about that, General. I'll get back to you on it."

Aster stared hard at Abbott for a protracted, silent moment. *Damned Washington lightweight! Gotta go ask Mama first. That little twerp's gotta be the ass who leaked those preliminary, unscrubbed options to the NSC! Just when we need real NSC horsepower out here, we get stuck with this peabrain!* Aster fumed, struggling to control a blinding urge to snap Abbott's neck. Blowing a fuse at Abbott would accomplish nothing, though, and he knew it. Instead, he calmly suggested that the NSC staffer work with Dillon to get a cleared-for-public-consumption story together as soon as possible. The general sat down and flicked a hand toward the pit, jaw muscles still twitching.

"And fourth," Androsin continued, "BOYD has corroborated a Russian tip about the antisatellite weapon's location. It's somewhere in the Dushanbe region. Okay, our job is to think *beyond* the ILM mission and the probable SOF follow-through," he said, shifting gears. "We'll examine likely outcomes, but we're also going to focus on what we need to prepare for next, after the satellite killers are out of business. We need to look hard at the potential ramifications of these spacecraft losses."

Androsin stole a glance at his watch, just as Aster reached for a vibrating handheld communicator. The general checked its tiny screen, then signaled that he needed five minutes. Androsin took the hint and added ten. Generals never took only five minutes.

"We'll take a fifteen-minute break, then devote about a half hour of work as individual teams. We'll reconvene as a group after that," the colonel concluded.

Androsin trailed the commander and his entourage from the center, reflecting on the battle staff watch commander's skepticism to Barb Leewon's explanation of the ILM mission. *That does it. Now's the time to make my pitch to the chief,* he decided. *We've been working at half throttle. If we're going to get ahead of real-world events, we'll need the full battle staff in here.*

6

A HOLE IN THE SKY

Genady released main-gear brakes, then smoothly pushed the throttle to its forward stop. In a familiar routine, born of long experience, his attention alternated between the concrete runway racing toward him and several instruments on a rather sparse cockpit panel. He eased the stick back, held it until the aircraft's long nose came up, then locked the attitude at a steep, book-prescribed angle. The runway quickly disappeared as the Myasishchev M-55 "Geophysika" climbed rapidly. The former Russian air force pilot raised the landing gear, banked to the south, then settled in for a long flight. The long-winged, slender aircraft continued to climb, gobbling thousand-foot altitude bands in seconds, headed for 60,000 feet above sea level.

Tajikistan was close by, especially when you cruised at 450 MPH, but today's mission profile was unusual. He'd be flying in a very large circle for several hours, keeping the reconfigured "lidar"—essentially a "laser radar"—and other sensors pointed toward the interior of that circle. Packed into a pod under the M-55's nose, the sensors were aimed forward and to the left today. On a typical M-55 mapping mission, they'd be pointed downward, soaking up ground-level environmental data, while other devices sampled the atmosphere.

The pilot trimmed his aircraft to climb at a stable, fuel-efficient rate and cross-checked navigation data to ensure he was headed for the right spot in the sky, though it was still far ahead. He turned his attention to a mission checklist strapped to his thigh and started a methodical in-flight survey of the craft's nose-pod payload. Everything was working, at least for now. He mentally corrected that flash of doubt, reflecting that the new American sensor package had proven to be much more reliable than the aging reconnaissance system he'd flown as an Air Force officer years ago. *Flying for hire . . . It is better than flying for war,* the pilot mused.

He was delighted to be in the air again. Sure, it had been an intense scramble to get the aircraft ready after months of downtime. But the American partner company, Intelligent Land Management, had made it very clear that a generous fee awaited Genady and his comrades, *if* these missions could be launched immediately. Bringing a new instrument payload configured specifically for this unusual tasking, ILM's president and CEO, Bryce Kameron, had made the long trip to Russia on a chartered Gulfstream 500 business jet. Unusual, for sure, and the high costs of such measures only underscored the urgency of today's top-priority mission. Kameron was emphatic that the flight be conducted in accordance with a very detailed plan. No pilot-initiated deviations, he had stressed.

Genady was sure the "consultant," who accompanied Mr. Kameron, was an American spy or military officer. Probably the latter, based on the consultant's demeanor and unmistakable quirks. Any short-haired man who aligned his shirt's buttons with the edge of a belt buckle and fly seam had been through some kind of military training. He smiled, eyes continuing to scan the M-55's instruments. Still on course.

Above, the sky's color had darkened considerably as the M-55 continued to climb. NATO had dubbed the M-55 "Mystic-B," but American Air Force pilots simply called it the "U-2-ski." That was a mistake, he reflected, prompting another smile beneath the oxygen mask. The M-55 climbed better than the American's Lockheed U-2 spyplane, and was much more forgiving to fly in the very-high-altitude, maximum-speed portion of the flight envelope that U-2 pilots called "coffin corner."

Finally, the M-55's nav system indicated he'd arrived in the search area, not far from Dushanbe, Tajikistan. Referring to the detailed checklist, he activated the lidar and infrared search-and-track systems, ensuring all sensors were pointed toward the aircraft's ten o'clock position. A six-inch video screen mounted temporarily in a shaded area, under the instrument panel's overhanging lip, showed a processed image being recorded on the hard-disk of a Panasonic Tough PC II secured beneath the ejection seat. For now, the display was blank. The pilot hoped he would see the telltale reflection of lidar energy Mr. Kameron had described, but that was unlikely. If it appeared, and Genady didn't catch it, a software-triggered alarm was supposed to sound in his helmet's earphones. Then he'd immediately adjust his flight pattern and start the triangulation process required to fix the position of a "hole in the atmosphere" Mr. Kameron had said to look for.

Genady wondered again exactly what could create such a "hole," and how it would ever remain open. His technical training to fly this airplane, one of the few Geophysikas still in operation, was extensive, but that training didn't help him visualize a "hole" in the air. Still, he watched the small, flat-panel screen carefully as he gently guided the aircraft, gloved fingertips moving the control stick almost imperceptibly.

The aircraft droned smoothly through the rarefied atmosphere at the lower edge of space. Genady could clearly see the Earth's curvature, its horizon a thin band of blue that gradually transitioned to dark purple, then black as one's eyes were drawn upward. A strange buzz in his earphones snapped the pilot's attention to the small screen. There! A blip of reflected laser light had been transformed by the video processor to a thick, pencil-like structure slanting across the display screen. He punched a position-ID button on the navigation panel, then quickly wrote the now-frozen coordinates on a form clipped to his kneeboard—just in case. *Never depend on a data-recording system . . .* The line of laser light remained on the screen for only a few seconds, then disappeared.

Suppressing his excitement, Genady quickly adjusted the M-55's flight pattern, reset a sweep-hand clock on the instrument panel, and ensured his aircraft was firmly established on its new heading. Exactly one minute later, he made another 120-degree heading change, then repeated the process several more times. He completed a couple of triangular patterns, ensuring the position of that brief lidar-exposed line of light was well documented.

The pilot remained on-site another hour, and was rewarded by a second streak of lidar-detected energy. This time, the data-processed representation of that beam remained a full five seconds, growing brighter with time. Fascinated by the bright, more-persistent line, Genady again flew the triangulation pattern with precision. *What could make a hole in the atmosphere? And how could it persist for so long?*

A fuel check told him it was time to turn off the pod's sensors and data recorder, and set a course back to the air base. He was confident Mr. Kameron would eventually tell him what created that fascinating hole. *A few glasses of vodka will loosen the American's tongue tonight.*

The handsome, well-dressed ILM president would be very pleased that Genady's first "special" mission had been so successful. Kameron had fully expected to fly six or more flights, and warned that the pilot might never see a blip of lidar-light reflected from the mysterious hole in the sky. Maybe the American would pay his Russian pilot a small bonus for his, Genady's, first-time good fortune.

11 APRIL/STRATCOM HEADQUARTERS; COMMANDER'S OFFICE

Androsin hovered outside Aster's office, mentally rehearsing his "elevator speech," a pitch so concise it could be presented in the span of an elevator ride. He'd only have a few minutes to convey his deep-seated conviction that the art of wargaming had matured and was ready to move into a new realm. It was now a viable tool for supporting *actual ongoing military operations* and aiding a commander's decision-making process in near-real-time. For too long, wargaming had been limited to strictly what-if scenarios, a bloodless way to explore "alternate futures." But now, faced with a "Pearl Harbor in space," Androsin knew they'd crossed into new territory. This game was dealing with rapidly unfolding, real-world situations, and its participants were *in* the fight, not merely future-gazing. What they were discovering was quickly being converted to war plans. *They were now fighting a real war inside the wargame center.* But he still had to convince his boss that the ideas-to-actions loop needed to be tightened.

Hope the boss is ready for this one, Androsin breathed as he entered Aster's outer office. *Even if he buys it, the battle staff is gonna have me for lunch. But it's the only way to work our way out of this space mess.* Annie smiled and waved the colonel inside.

Androsin entered, then waited while Aster surveyed a large, wall-mounted battle screen. "Marvelous technology, Jim," Aster mused. "But technology isn't serving us very well right now, is it?"

"That's right, sir. 'Observers use awareness and decision makers learn from actions. Winners are a blend of all elements.'"

"And your point, Colonel?" Aster's clipped reply had an uncharacteristic edge. Events were taking a toll on the general's patience.

"Sir, you've said we—America—really screwed up the space business in the nineties and early twenty-first century. Disestablishing U.S. Space Command and merging it here with STRATCOM changed STRAT from strictly a user of 'space' to both a user and provider of space capabilities. So, my point, sir, is this: It's not the technology, never has been. It's *how* we make decisions, in peace *and* war. We don't pull all the elements together, resulting in a poor, fairly undisciplined decision-making process despite the command structure."

Aster shot his wargaming chief a hard look. "Okay, Jim. You've got a burr under your saddle. You telling me my battle staff doesn't know how to plan and make decisions?"

"No, sir. It's not just about the battle staff; it's the whole kit-and-caboodle," Androsin said, a bit nervously. "General, we can improve both the wargaming process and the battle staff's planning. But to get there, we need to bring the battle staff *into* the wargame. Integrate them into the game's teams, then designate a core element to develop and report findings and recommendations directly to you." Androsin paused, waiting for a reaction.

Aster crossed his arms and continued to stare at the big screen. "I thought we already had a battle staff element in the current game."

"True, but they aren't the decision makers, just the gofers. They have no real stake in the outcome. I'm suggesting we bring in the entire battle staff," Androsin continued. "Merge the operation. Fight the war right here, *right now,* in the wargame center. In other words, make the war*game* the *real* war because it already is."

Aster considered the possibilities, but appeared skeptical. "How do we do this?" he asked, encouraging Androsin to press his point home.

"In this way, sir, everybody will be better able to 'observe and orient' the battle space. Not only in military terms, but in the broader, geopolitical context of the moment. We already have a solid group of interagency types in this wargame— plus commercial space guys and those fifty-pound heads from academia—but we still need to immerse the battle staff into the game's routine. That way, our guys can learn first hand what our allies are thinking before they pull the trigger, not after. They will see how our potential adversaries might act, as seen through Admiral Lee's threat scenarios. They'll get a better appreciation for things like cultural factors and how events are shaped by real-time environments. And they'll see that our analyses are rigorous processes—especially as we synthesize BOYD's outputs. In short, the staff will gain better, deeper insights to wargaming's proven methodology for testing the consequences of decisions in a no-harm, no-foul laboratory *as they actually make the warplanning decisions*."

Aster was nodding slowly. "Jim, I think you're onto something, but help me understand the advantages and downsides. What exactly do we gain by embedding our battle staff with the gamers?"

"Two things: time and the quality of decisions, sir. As you know, we always have a postgame hotwash that summarizes findings and recommendations for you and your senior staff. Then we give the battle staff a distilled version of the same thing, but with your commander's intent added. It acts on that distilled version—maybe 'sanitized' is a better term—plus your direction on how to proceed. I'm just saying, let's bring the staff into the wargame so it's part of the upfront, total experience as a way to shorten our overall timelines.

"Right now, events are unfolding so fast that we need to do a better job of getting inside the other guy's OODA Loop," he stressed. "We can't afford the after-action hotwash session. By then, it's too late. This space war is happening right now, and we're fighting it in the gaming center every day. Our Blue and Red teams are giving us that few-percent edge needed to win. If we put 'em in the middle of the wargame thought process, our battle staffers can very quickly turn game outcomes into executable actions that have both tactical *and* strategic impacts." Androsin studied Aster's expression. The general was known as a skeptic, but he was against the wall here, because events were unfolding so fast. And the colonel knew it.

"John Boyd's OODA Loop again, huh? Makes sense, though." Aster was pacing, arms folded, staring at the office's carpet. He considered the breakthrough strategic thinking done decades earlier by Colonel John Boyd, a brash, innovative fighter pilot. Boyd's theory—*get inside your adversary's decision-making process; out-strategize and head him off at the pass before he even figured out there was a pass ahead*—was clearly applicable to the current, rapidly degrading space situation. *Damn sure does make sense,* the general concluded. "Okay, Jim. Sold. Let's make it happen. Anything else?"

"Sir, Admiral Lee's a huge proponent of Sun Tzu and the strategic thought process, as you know. Well, the admiral's already thinking well beyond this DEADSATS II wargame, and he's moved his Red team into new territory. He's got some pretty scary scenarios laid out, sir. It occurred to me that, with our battle staff directly tied in, we could really lean forward and start doing some real-world strategic planning based on Lee's scenarios. If some of them start to unfold in real time, we'll be ready. Sir, I suggest the war's already on and Admiral Lee's fighting it, even before each battle happens."

"The ol' admiral is one smart son-of-a-bitch, all right. After I was confirmed for this job," Aster reflected, "he was all over me, adamant that I read *The Art of War*—the whole thing, cover to cover. I thought I'd had about all of that I needed back in war college, but the old man said 'no way.' He kept badgering, insisting I had to fully understand the concept of *cheng* and *ch'i,* the traditional and the unexpected, the balance of opposites, especially asymmetries that are organic to the practice of warfare. That's what we're looking at right now, so maybe you're right.

"We damned sure need to get some space-capability regeneration going, too! So, let's get our heads out of the traditional-thinking box and push both our gamers and battle staff out ahead of the unexpected." Aster paused, then added, "You think the other gamers'll buy into this, as well as what Lee's doing?"

"I'll work on them, sir. We've allowed the ol' leadership axiom of 'let chaos reign' go long enough. It's time for the other half: 'then rein in the chaos.' Admiral Lee can be a big help on that score. And with General Forester cochairing the main Blue team, I think we'll get some good options his ops guys can act on, and soon."

"Make it happen, Jim. I'll prep the battle staff and get Buzz

Sawyer's Space Command people moving on some serious reconstitution efforts. Regardless of what our people come up with in DEADSATS II, we've got to restore some of these space capabilities. And if we lose any of our 'Battlestar Galactica' platforms, we're really screwed."

Androsin left, headed for the wargaming center. His stomach was churning, stirred by both excited anticipation and dread. *Holy crap! He bought the whole enchilada! What the hell did I just do? And "Battlestar Galactica," has to be our behemoth imaging and sigint satellites! Maybe the huge Milstar comsats, too. They're supposed to survive a nuclear blast and keep on ticking, though. If we lose some of those, we'll damn sure be blind and deaf! Aw shit . . . !*

11 APRIL/DUSHANBE, TAJIKISTAN

The chill of an Eastern European night seemed to spread throughout the small bungalow Alexi shared with his wife and daughter. Though it was early spring, the damp cold felt more like November. Through a sleep-deprived haze, Alexi was again struck by the tremendous impact of Nadia's efficiency, resourcefulness, and technical acumen. He looked up from his computer screen, feeling a father's intense pride swell from deep inside. Through the window of a small, cramped control room, his tired eyes followed the beautiful, brilliant girl who was his—*his*—only daughter. *It is she who made this breakthrough possible,* he reflected. He leaned back to rest briefly, relishing the moment, watching her work in the basement below. That flash of pride helped clear the incessant anxiety that plagued him, a chronic affliction he attributed to Domingo's periodic, thinly veiled threats.

The dank basement was crammed with an unlikely mix of hardware. Delicate, sophisticated traveling-wave tubes and waveguides seemed to float only a few meters above the hulks of several refrigerator-sized generators. Cables as thick as a man's forearm snaked across the floor and up a far wall, paralleling a bank of machined-metal waveguides and a bundle of fiber-optic cables that stretched to, then through, a paint-peeled ceiling.

Nadia seemed to sense her father's eyes following her. She turned and flashed that captivating, stunning smile that had broken many Russian men's hearts, Alexi knew. Suddenly, a fit of coughing caused her to bend, the toes-deep cough racking her slender frame. A small outstretched hand gripped an iron support. She was still too weak. It pained Alexi to see her working in the damp, chilly basement for hours at a time. But she'd insisted, and he clearly needed her skills and persuasive powers. Now more than ever.

Alexi initiated yet another system-architecture analysis via the new computer Nadia had miraculously acquired. He let his mind relax, knowing that clear thought was the most reliable avenue to the answers he needed now. Was there anything he'd missed? Was the system truly ready for "de beeg show," as Domingo had said? It didn't seem possible, at least so soon. But Nadia had found a most ingenious way to squeeze even more power from the novel weapon system in that basement. By rewiring it, then precisely controlling the multiple phases of massive electrical generators, she had avoided a need to find, buy, deliver, and install the extra power units Alexi had calculated he would need. *Brilliant! Absolutely brilliant!*

Yes, he'd been extremely dubious, despite Nadia's conviction and assurances that her design would work. He worried about Domingo's reaction, should the new system not only fail, but also delay the modifications Alexi feared must be made. He shouldn't have, though. Nadia's smile and pedestrian explanations had convinced the pudgy Colombian that pursuing her concept was the fastest way to achieve Domingo's goal. Not that the brute understood a single electrical-engineering word she said, Alexi knew.

A coy smile and melodic words were more convincing than her solid engineering, he thought. But no matter. She'd backed her words with very rapid, intense action, and the system was now ready. Yesterday's "smoke test," as she called it—her American university slang still disturbed him—had been flawless. She needed only to complete some "tweaking" of the multigenerator phasing to ensure peak power was available when Alexi demanded it.

The Russian scientist smiled, recalling how nearly ecstatic Domingo had been yesterday, watching Alexi fire the low-power acquisition laser, then the laser/maser combination. They'd painted an American KH-15 "Hunter" spy satellite, but Alexi knew he had not hit the spacecraft hard enough to damage it. He wasn't about to risk frying his maser and upgraded power-generation suite for that initial test of Nadia's new system configuration. But it was enough for Domingo. He had hugged Nadia and tried to kiss her. Thankfully, the girl had been nimble enough to escape the animal's clutches, while leading him to believe she enjoyed the attention.

Tomorrow, the Colombian would return, though. And he would demand to see "de beeg show," an attack on two of America's icons. Alexi had cautioned that it may be too early, that the modifications Nadia and he had made might not deliver the power levels he'd calculated as necessary. But Nadia had been right all along. The system, indeed, was ready, with megajoules of energy poised for precise delivery. Alexi tingled with anticipation. Only a few hours until Nadia, he, and his ailing nuclear-scientist wife would be wealthy beyond their greatest hope of hopes. He could almost hear the waves lapping on the warm, white-sand Caribbean beaches of the Dutch Antilles.

11 APRIL/SHIR KHAN, AFGHANISTAN

Captain Dirk Baldwin didn't like waiting. Never did. *Get this friggin' show on the road, or get us back to the damned base,* he groused to himself. It was cold out here in the northern boondocks of Afghanistan, close to the Tajikistan border, where decidedly unfriendly Pashtun and Taliban renegades still roamed. His guys were restless and he, their team commander, didn't know what the hell was going on. *Mushrooms! Keep us in the dark and feed . . .*

"Sir!" Baldwin's personal bitch session was interrupted by his communications chief, a young specialist wearing a fresh Ranger tab on his shoulder. He handed the officer a scarred Iridium satellite-phone handset. Why they were stuck with these ancient, brick-size jokes was but one more small reason for Baldwin's funk. His unit had used the new lightweight, low-power Excalibur phones during a joint exercise with the Afghan and Pakistani armies only two months ago, and the comm systems had performed beautifully. So, why'd the task order from CENTCOM headquarters specifically dictate the use of old Iridium shit for this op? Didn't make sense.

Baldwin grabbed the phone and checked in, noting the annoying delay typical of a voice-scrambler somewhere in the Iridium system. Nothing like the near-instant Excalibur units had delivered. He listened, jotted a few notes in a field notebook, checked his watch and promised to "stand by until you give us the 'go.' . . . Got it, sir." He handed the Iridium set back and shook his head in answer to the specialist's silent question. "Nope. Not tonight. But CENTCOM said we'd better keep our boots on anyway. Something's about to pop." He headed for the hangar to let the Air Force Special Operations Command CV-22 Osprey crew know they were down for the night. Another no-go.

11 APRIL/STRATCOM HEADQUARTERS/INTELLIGENCE SECTION

"Right . . . there. That's gotta be it," the sergeant said, jamming a finger at an oversized photo. His civilian supervisor slid a jeweler's loupe, an old-fashioned magnifying glass of sorts, across the photo, put his eye against the lens, and studied the spot indicated. The "supe" didn't trust high-resolution digital images on a computer screen, regardless of how the young geek-wizards sliced and diced them, squeezing an incredible amount of information from a single image. He still wanted to see the no-shit color photo. He knew he was looking at a high-res printout, not developed film, but it didn't matter. Old habits died hard.

"Yeah, something's there, all right," he agreed. "A residential building, but it has some kind of greenhouse structure on the roof. Pull up the hyperspectral-sat data, then overlay it with infrared data from Global Hawk, and correlate the Hawk's position with that ILM stuff we got yesterday." With their incredible data-fusion capabilities, computers had their place, he admitted silently.

Seconds later, he knew the sergeant's assessment had been bang-on. Everything lined up perfectly on the computer display. That M-55-acquired ILM data had reduced the uncertainty-box to about a city block–sized area on the outskirts of Dushanbe. And unmanned RQ-4 Global Hawk aircraft platforms, which had mapped a much larger area for several days, in both the visible-light spectrum as well as the infrared, had detected an unusual amount of heat periodically emanating from that same building. Spysat hyperspec data showed no sign whatsoever of foliage, cuttings, or anything organic around that "greenhouse" on the roof, either.

"Have ops task an RQ-4, a U-2, satellite—whatever we've got in that region—to zero in on that building. Watch it like a hawk and don't blink," the supervisor ordered as he rolled up the large print, then twisted a rubber band around the paper tube. "Summarize all that info and get it onto the SIPRNET. There're a lot of heavy hitters out there who need this stuff, so move it pronto-like. I'm running a hard copy over to the command section." He headed for the door, then halted. "Oh! And make sure all that correlated data gets sent up to Jill Bock so she can integrate it into her magic databases."

11 APRIL/STRATCOM WARGAMING OPERATIONS CENTER

Jill watched the full STRATCOM battle staff shoehorn itself into the center, amused at how people tended to get territorial after only a few days. The posturing around what few fixed-position gaming stations were available was a study in personality types. Orderly, spit-shined military battle staffers were spread among a diverse collection of wargamers. Androsin had offered a pithy description of the integration process. The same routine always played out among soldiers boarding a truck, sailors on a ship, or airmen building their personal "nests" on a transport aircraft. *Ya just gotta let 'em get through all the dog sniffin' before they get down to business,* he'd said. So true.

Aster swept into the wargaming ops center, ordered the room "at ease," then addressed the battle staff and wargamers. "We're homing in on the location of the anti-sat electromagnetic weapon. Things will move pretty fast now, but you'll be fed the latest info as we get it. Keep working on the who and why, though. We can't fall into the old drinking-our-own-bathwater trap, so keep our options open, and keep *all* Red team possibilities on the table. There's too much at stake here, and we can't afford to zero-in on the wrong players.

"There *will* be responses, folks!" he added. "The president made that clear in his TV address last night. It's our job to make sure we're responding to the right bad guy, or guys. I hope you staff folks brought your sleeping bags. We could be here a while."

Aster stepped from the pit, tossed his laser pointer to Androsin, and waved a "carry on" to the group before its uniforms could jump to attention as the general departed. Androsin activated the big-screen's display and pointed to a table of figures.

"Let's review the latest fallout of all these satellite losses," he said. "A lot is happening, and happening quickly. Boeing's put a hold on more than thirty commercial satellites now in the pre-launch pipeline while their engineers review the -601 series in detail. The most immediate impact is a delay in the launch of JCSat-10, a Japan Satellite Systems direct-to-home TV bird. Boeing is definitely feeling the pressure. Its stock dropped about ten percent after the TVBS-1 failure, and analysts say it's still edging down.

"But the big news," Androsin continued, "has erupted in the insurance industry. A consortium of insurers has announced what they call a 'formal review of premiums' on all Boeing-built satellites. That usually means they'll jack up premiums while simultaneously cutting their exposure to further losses. In other words, about thirty Boeing customers will probably have to ante up again for higher-cost insurance, probably before launch, and/or have to assume more risk themselves."

The Boeing rep raised her hand, akin to a schoolchild. Androsin pointed to her. "Yes, Mitch."

"Audrey Mitchell from Boeing, Colonel," she said. "I'd just like to point out that this string of HS-601 failures still has few common elements, but we know a bit more now. Our engineers have found that, on three of the spacecraft, a brief, unexplainable rise in electrical current occurred between the control processor and a power distribution module supplying that processor with fifty volts. That caused irreparable damage on all three satellites."

"Do they really know why, though? What's the causal mechanism?" asked a ViaSat vice president, another user of Boeing's ubiquitous spacecraft "bus," the basic space platform that carried and supplied power for ViaSat-built electronic payloads. A satellite "bus" was analogous to a commercial aircraft's basic airframe, which could be configured in a variety of ways to suit different airline customers.

"They've pretty much confirmed the CIA's theory that some kind of intense, powerful, electromagnetic pulse is the root cause," Mitchell said. "At the moment, our people only know the end result was a power surge that caused onboard failures. But my point right now is this: Boeing's stock has dropped sharply, as the colonel indicated earlier, but the biggest impact has been on future spacecraft orders. A number of customers have frozen negotiations, and several have already gone to our competitors. Those factors are driving Boeing's credit ratings down, which increases the cost of capital we now have to raise. In other words, these losses are having a very rapid, very adverse financial impact on my company. And we're one of the biggest satellite builders in the world today. I think we need to ask ourselves: Are these spacecraft losses triggering a domino-like reaction that could devastate the entire commercial space industry? And it's spreading beyond the U.S."

She let that sink in, lest any of Boeing's competitors get the idea they might be gaining an advantage. This wasn't a Boeing problem, not by a long shot.

"And there's another issue, maybe way out on the horizon, but out there nevertheless," she continued. "We all know the Russians and Chinese have been undercutting our launch-vehicle prices for years. We also know that the dollar has been the currency of choice for transactions throughout the global space industry. However, if U.S. companies have to cover higher insurance premiums and costs of capital, could the euro become the currency of choice? This happened in the petroleum industry a few years ago. What would be the impact on our industry? Something to keep in mind as we look for bad-guy motivations, I suggest."

Mitchell had been the sole industry representative to approach STRATCOM on its own, raising the same concerns the DEADSATS I wargame had triggered a year ago. Clearly, the aerospace giant's executives were worried that a catastrophic meltdown of the U.S. commercial space industry could occur. Androsin exchanged a quick look with Lieutenant General Dave Forester. The ops director's nod confirmed that he had caught the same, perhaps unintended, message in Mitchell's comments. There was something significant about all this financial gibberish, and they'd have to get some outside help to see what it all meant.

Damn, who'd ever think STRATCOM would need a friggin' MBA to sort out space financials? Androsin mused.

11 APRIL/STRATCOM HEADQUARTERS/COMMANDER'S OFFICE

"What's your level of confidence on this?" Aster asked, leaning over the large photo. The tiny damned thing just looked like a house to him.

"Over ninety percent, general," the civilian intel supervisor said. "We've cross-checked it across the spectrum, and all data point to that site. Something at that location is periodically producing a hell of a lot of energy, and the beam ILM's

M-55 detected yesterday came from that general area. And the hyperspectral signature of that doghouse on the roof confirms it ain't no greenhouse. It may look like one, but it isn't. Short of putting a body in there, that's as confident as intel gets."

Aster appeared to be studying the image, but was really sorting pros and cons. He'd have to take the risk. "Thanks. Tell your guys they've done a hell of a good job!" He escorted the intel chief from his office, then turned to Annie. "Would you round up Dave Forester, please? I need him and his mission-planning guys in here ASAP. And get me the SecDef."

By phone, Aster quickly outlined the multiagency intel findings, then asked Secretary of Defense Hurlburt for a confirmation that he was, indeed, cleared to launch an SOF mission against the Dushanbe site.

"Hell, you already had a 'go' clearance, Howard. Just go *do* it," Hurlburt growled. "Make sure you keep CENTCOM in the loop, though. I'll let State and the White House know we have an op underway out there. The president will get all loosey-goosey about 'what-ifs,' but I'll handle that. Just make *damned* sure we don't screw this up! Uhh . . . do we have any indigenous involved?"

"Yes, sir. One Tajik liaison officer's with our SOF team. They're all camped up near the Tajik/Afghan border, stayin' clear of the locals, but primed and cocked. Soon as my folks update the plan, we'll launch 'em."

11 APRIL/STRATCOM WARGAMING CENTER

It was getting late, and nerves were fraying, Jill noticed while surveying the small groups huddled here and there around the center. Her eyes hurt from staring at the screen, scouring BOYD, SIVTRIX, multiple intel feeds, even the online financial press reports for a good fourteen hours. But they were homing in on a solution. She could feel it. Androsin had just whispered to her, "We've found the site; the SOF op is a go."

But he had *not* passed that tidbit along to the wargamers and what battle staff remained, she'd noticed.

A sudden outburst attracted glances as NSC's Abbott exclaimed, "Good God, man! You can't just launch an air strike on civilians inside another nation! That's an act of war!" The pudgy NSC rep's face was sweating and getting redder. A stereotypical inside-the-Beltway animal who had cut his bureaucratic teeth as a State Department staffer before being detailed to the NSC, Abbott had always managed to stay far removed from the realities of war and national defense. He'd been nurtured for years in the cocoon of academic idealism, smug in the knowledge that the superior intellects of his kind had allowed him to quickly rise above the disgusting, warmongering world of the Pentagon's redneck Neanderthals. But now he was nose to nose with one.

"So, what the hell do you recommend?" an equally redfaced Navy admiral asked sarcastically. Vice Admiral Ted Fraser, Aster's three-star Navy deputy commander of STRATCOM, had agreed to drop by for this particular round, sitting in for his boss. Fraser was a former SEAL and had practically written the book on modern covert operations. He had no idea that an SOF mission was already under way in Tajikistan, but had casually mentioned such an option was on the table, as was a B-2 stealth bomber precision airstrike on the beam-weapon site in Dushanbe.

"Why don't you just whip out that li'l ol' cell phone and call one of your State Department weenie buddies and politely ask them to get their butts over to Tajik-land and 'negotiate' with those boys, *son!*" Fraser had dropped into his native Texan dialect, and was losing control of that infamous temper, Androsin noted from across the room. Should he step in and maybe help preserve what remained of the admiral's stellar career?

Too late, he saw, picking his way through the amphitheater's barriers. Abbott was fuming now. "You son-of-a-bitchin' military fools! Only have one answer to every world problem, don't you? Go in with guns a-blazin' and blow up all the bad guys, even when you're not even sure *who* the bad guys are!

Remember Fallujah? The Afghanistan border? Pakistani villages? What the hell do you care? It's all about body counts! Just stir up a little action and get another li'l ol' star!" The snotty young NSC rep had committed a cardinal offense, mimicking Fraser's accent. Abbott was toast. Androsin had seen Fraser in action before and knew what was coming.

"Well, if you political desk-jockeys had a single red-white-and-blue American hair up your fat li'l ol' asses, we'd take care of this satellite-bashin' right *now*!" Fraser bellowed, standing up so fast his chair careened away. He was towering over the now shrinking, suddenly very short and fat Abbott, body language suggesting that the young bureaucrat was about to be handed his head.

"You politico farts demand we protect billions of dollars' worth of space stuff, but you keep cutting defense budgets to pay for stupid back-home pork projects!" Fraser shouted. "Then you tie our hands with that naive, idiotic 'no weapons in space' bullshit! How the hell are we supposed to protect and defend this nation with assholes like—"

"Knock it off, Ted!" Aster barked from the amphitheater's rear doorway. He'd returned to the center unannounced, planning to pop in and quietly monitor the intriguing insurance-related discussion. Androsin had arrived at the almost-battle scene at the same time, grabbing the admiral's rock-hard biceps. An Army battle staff major also had Abbott firmly in tow, edging him away from the ex-SEAL admiral. Aster appeared beside his deputy, grasping the admiral's other arm. Androsin stepped between Fraser and Abbott, one hand aimed at the blocky senior officer's chest.

Fraser took a deep breath. "Sorry, boss. Guess I . . . I lost it."

"Damn right you did," Aster hissed in a barely audible whisper. "Don't forget who that little bastard is, Ted! Get a grip, man!"

The room had become deathly silent. The assembled civilians held their breaths, eyes wide at having just witnessed two powerful government servants practically come to blows.

Androsin quickly took charge of the standoff, even though

several others outranked him. "Let me take a coupla minutes to kinda summarize what's going on here," he spoke rapidly and informally. The major kept a firm hand on Abbott's shoulder, having pushed the red-faced NSC rep into one of the thirty-some seats fixed to the tiered room's floor. "You've identified that potentially disruptive signals are being beamed from Dushanbe. These have caused serious problems with at least nine, maybe ten, on-orbit U.S. commercial and national security satellites. Diplomatic efforts have been unsuccessful in terminating the signals, while creating something of an international crisis between the U.S. and Tajikistan," the colonel said, his words a staccato stream.

"One of the options brought up for consideration is launching a stealth bomber airstrike to take out the site where the signals are originating. That idea was . . . uh . . . let's just say it wasn't well received by our National Security Council representative," he said, glancing down at Abbott. The now-sweating NSC staffer still glared defiantly at Fraser. Androsin glanced at Aster, who shook his head, almost imperceptibly. The general, for whatever reason, wasn't ready to reveal the SOF operation in progress.

Androsin descended several steps to the pit and activated the room's large central screen. "Here's the official Tajik position: 'No sanctioned activity' that could explain the signals we've detected. Within the last hour, though, we've isolated the signals' source to a specific residential area. But we have no—repeat no—indication that a Tajik military or other government facility is within that area of the city.

"The bottom line is this: In the absence of stranglehold proof, a diplomatic solution is out. Our attempts along that line have only strained relations with the Tajik government. We're certain those satellites are being attacked from this area, but we're into a realm that has no solid political or legal precedent. Satellite-bashing by a nongovernment entity, *if* that's what we're dealing with, just isn't addressed by international law."

"I say a covert, rapid strike would put the damned Tajiks

out of business and send a warning to every other bandit that ya don't screw with the U.S. bull or ya get the horn!" Fraser said, still steaming.

Aster had heard enough. "Let's get back on track, people," he snapped, then changed his mind. "Hell, maybe we need a break. Five minutes. No more. And when we come back, I want an answer to who's behind these attacks."

7

FOLLOW THE MONEY

Admiral Fraser had left the center, scowling at Abbott, who was sitting on the sidelines, punching numbers into his comm device. Air in the amphitheater remained thick with tension.

"I want to hear from the analysis group," Aster commanded, imposing raw discipline on the mixed group of civilians and his military battle group. "Jack, Rob, George, what've you come up with?"

"General, we think we may have found a motivation for the attacks," the CIA's Tanner said. "And maybe a way to eliminate some of the potential suspects to home in on the right bad guys. We're still checking, and this is a little tenuous, but it goes like this. Given the intel pieces we've received, plus data from the environmental outfit that flew the M-55 'U-2-ski' for us, and an interim report from an ongoing Treasury Department and Justice Department anti–money laundering task force in California, we looked at who might stand to gain from commercial satellite losses.

"Government-sponsored action just didn't make sense to us. Harassment? Maybe. But we just couldn't find a solid reason for the Tajiks to go after our satellites. What's to be gained from pissing off companies that they're desperately trying to do business with? But then . . ." Tanner hesitated, tapping keys at his station to bring up a series of bulleted items on a screen behind Androsin. The display was blanketed with lines and arrows.

"We're still in the early stages of this, and it may be hard to follow, but please walk through this with me. When you look at all the possible players, and how they might benefit from TransAmSat and Boeing and Pantera and God knows who else losing multimillion-dollar satellites, you find only one intersecting party."

A stylus-drawn pointer moved to the confluence of several multicolored arrows. "The insurance companies."

"Now, wait just a damned minute!" an angry voice bellowed. Tanner turned to face Merle Beatty, an elderly, balding, and slim man with a huge white mustache, who had jumped to his feet. Tanner noticed that the man's mouth was completely hidden; only the brush above it moved when he spoke. Until now, he'd remained virtually invisible, appearing only mildly interested in the wargame. "I represent the Association of Space Insurers, which includes most of the companies that provide coverage for your satellites," he explained, turning to the crowd of at least fifty people.

"We all lose money when *you* folks lose a bird. Now, why would we want to cause problems like that? And how in *the* hell could we even generate an attack like that? We're not a bunch of engineers, in case you haven't checked your agent's resumé recently," Beatty deadpanned.

"I know, Merle," the CIA's Tanner nodded in agreement. "But stay with me for just a minute and this'll start to make sense, I hope." He turned back to the keyboard and called up a new display. "As these data show, almost all the major insurers have raised their rates on commercial satellites, and reduced the risks they'll insure, in the three weeks since the TransAmSat and Excalibur problems arose. Now, Merle, you'll see that only a few of these companies had to pay out for those losses. But *all* companies raised their rates and cut coverage. It looks to us like the net result might be a profit, not an industry-wide loss."

Beatty's prominent brow furrowed as he twisted the corner

of his moustache, nodding slowly. "Go on, I'm still on board," he said.

"Let me ask you, if just one insurance company were behind all the satellite losses, would it go under?" Tanner asked.

"I doubt it," Beatty said. "I can't get into why—industry proprietary info, you know," he said, sounding a bit pompous, Aster thought. "Just assume that there are protections in place."

"Pardon an old soldier's ignorance of the business world, but what the hell are you talking about, Tanner?" "Matt" Dillon asked. He was still tight-jawed from his less than sterling press-briefing experience, Aster noted. The reporters had torn Dillon apart, grilling him about spiraling spacecraft losses, and he was still stinging.

"Maybe I can help here, George," Beatty interrupted. Tanner waved him on. "If I follow the case example that George's team came up with, one insurance company might have an incentive to cause problems for some other company's satellites, just for competitive reasons. Maybe try to drive the other guy out of business. Hell, I don't know. But what George doesn't know is that this doesn't make sense. Sure, the hard-hit company would have to pay out, but—and I probably shouldn't be saying any of this—if that company is part of our association, their risk is minimal. We've all agreed to help cover the others' losses, assuming one company is hit especially hard by multiple failures. It's sort of an industry self-protection thing," he added, almost embarrassed by revealing an insider secret. "Because if an insurance company goes under, our assets have to be stretched thinner to cover additional companies. And there's only so much money to go around."

"But where's the big gain? Why even bother taking down a bunch of satellites?" Dillon pressed.

"That's where we've gotten stuck," Tanner said. "Surely, even with that self-protection, there's not enough money involved to warrant disabling a flock of spacecraft, is there?"

Beatty shook his head. "Not that I can see. Usually, the company that covers the lost birds only gets back part of its losses from our association. All the other members' contributions do is cut those losses a bit. The system's designed to provide just enough to keep an outfit from going under. Nobody's getting rich, that's for damned sure."

Aster had listened carefully, absorbing but not judging what was being said. Yet something had clicked. "George, I think this avenue is worth pursuing. There's money flowing here: money coming in and money going out. Make sure you tap your FBI buddies, the white-collar crime guys. Or do you have any of those?" Aster smiled, unable to resist a jab at the ongoing FBI-CIA cooperation foibles. "Maybe go back to ping that ongoing task force investigation to see whether they have anything on these insurance guys that might help us. Follow the money, and it'll usually lead to the crooks. Let's be sure everybody who can help us is read into this wargame."

Androsin smiled. All the hard work he and his game developers had put into this was starting to pay off. They were getting closer to answers.

12 APRIL/DUSHANBE, TAJIKISTAN

Domingo's BMW bounced through a pothole, unseen in the dark street, and screeched to a halt. The fat man jumped out and waddled at a half trot to the nondescript house, then pounded on its door. Alexi appeared momentarily, hair mussed, a ragged housecoat draped over his shoulders. "Mr. Domingo! What . . . ?"

"Lexi, get your skinny ass into that control room. We've got some bad shit going down," Domingo hissed, shoving the Russian back into the room and kicking the door shut. Wide-eyed, Alexi stumbled, stunned. The Colombian's smiling, playboy persona had vanished. The man before him was now mean and dangerous.

" 'Going down'? Down where? I don't . . ." Alexi had backed into the darkened, sparsely furnished living area, confused. Domingo started talking rapidly, yet quietly, enunciating clearly, his Colombian playboy accent conspicuously absent.

"Look. A very well-paid contact in the Tajik defense ministry just called my hotel and disturbed my . . . He said the Americans are on to us. They may be headed this way now! You get Nadia out of bed, get that thing fired up and hit those targets we ordered earlier. *Now!* You understand?" Domingo's face was inches from Alexi, his pudgy right index finger repeatedly jabbing the scientist's chest, breath reeking of alcohol.

Alexi felt his bladder trying to embarrass him. He gulped and nodded, then turned and hurried down the hall. He was so scared his legs threatened to collapse.

Minutes later, Domingo hovered over Alexi's shoulder as the scientist's hands flew across a computer keyboard. Nadia, dressed in a baggy sweatshirt, jeans, and sneakers, rapidly flipped switches. Her eyes flashed between a three-ring binder-mounted checklist in her left hand and a control panel secured to the small room's side wall.

"Okay, Father. System is powered and primed. Charging capacitor banks," she clipped.

Alexi's mind, still dull from being awakened so suddenly, struggled to make sense of data on the glowing flat-panel screen. "I need radar feed, Nadia!"

The girl squatted before a second computer jammed into a vertical rack of equipment, scanning its memory for a certain radar file. Thanks to a dear young Moscow-based officer, who had professed undying love for her, Nadia had unofficial access to the Russian air force's raw satellite-tracking radar database. Bouncing those data against the American "space catalog" she had hacked into through a Texas university's server, any space-based targets of interest could easily be located. Super-classified American satellites that did not appear in the open-source space catalog showed up in the Russian radar data, making them easy to locate. Their orbital parameters also were excellent clues about the platforms' probable missions.

"I have it," Alexi breathed, almost reverently. *Thanks be to God! We may live through the night . . .* he flashed, only too aware of that chrome-steel automatic under Domingo's rumpled coat. Actually, finding the same Hunter imaging-intelligence platform that he and Nadia had radar-painted the day before, the one Domingo's Iranian "sponsors" had found so compelling, wasn't too difficult. Nadia's clever comparison technique worked every time. Now he had a graphic of the spysat's elongated orbit on screen. Its current position was hardly optimal, when viewed from Dushanbe, but it *was* within lethal range. A trackball moved a small white box over the satellite's icon. Alexi left-clicked a button on the track ball's base and the box shrunk, surrounded, then followed the spacecraft's slow movement across the display.

"Ready for ranging," he said.

Nadia stood and pressed a button, opening a port in the fiberglass doghouse structure's roof three meters above their heads. She visualized the exotic, gun-like apparatus now aimed through that port. Pressing a second button fired the low-power acquisition laser. Alexi prayed for a clear sky. Thick clouds could attenuate the laser's energy, even though its frequency had been chosen to accommodate some cloudiness. Again relieved, he saw the white box on his screen flash and turn red. "Locked on," he said, too softly. Nadia glanced at him, worried by the stress in his voice. She raised a plastic switch guard and positioned a thumb under the toggle switch.

"The system is ready, Father," she said in Russian, trying to steady her own heartbeat, praying the new "tweaked" design would not fail.

"Speak English! You *will* speak English!" Domingo shouted.

"Reaaaddy . . . *fire!*" Alexi squeaked. Nadia thumbed the switch and held it, eyes locked on a power meter. She noted the expected needle drop, then released the spring-mounted switch. The meter's needle slowly started rising. "Beam away. Recovering; rate normal," she intoned flatly. Domingo's outburst had set her heart racing, but she tried to avoid showing a hint of fear. Then she choked and started coughing uncontrollably.

Meanwhile, a second, more powerful laser blasted from the rooftop and bored an ionized hole through the thick lower atmosphere, spreading slightly as it paralleled the low-power range-and-acquistion/tracking laser beam now glued to its target, the unsuspecting Hunter spysat. A microsecond later, a massive electromagnetic pulse raced along that ionized-air "waveguide" cut by the high-power laser beam. Bursting into the expansive emptiness of space, the maser pulse's invisible wavefront blindly followed the laser beam's path.

Hundreds of miles overhead, yet only 32 degrees above the horizon when viewed from Dushanbe, one of America's most secret satellites was slammed by a powerful EM pulse. On-board sensors immediately registered what they were programmed to detect: the electromagnetic pulse of a nuclear blast. Solid-state electronic "latches" triggered, instantly reconfiguring circuits to a self-protection mode designed and fabricated in tiny, intricate patches of silicon more than a decade earlier. Theoretically, the K-15 space platform was "safed," prepared to ride out a nuclear attack. Then it would go back to work, sending extremely high-resolution digital images back to ground stations.

But time, exposure to billions of cosmic rays and the ultra-cold of space had taken their toll on the aging spacecraft's bus and the payload's integrated circuits. A critical power supply surged, then failed. Its backup system tried to assume the electrical load, but couldn't; another preprogrammed sensor "shed" nonessential electrical functions, and the satellite resumed operation, although in a degraded mode. Its large-aperture lens still stared earthward, but damaged attitude-stabilization circuits could no longer keep the silent, probing eye pointed precisely where its masters ordered. Within days, the craft would be all but useless, a billion-dollar derelict staring into deep space, spinning slowly, aimlessly.

Thousands of miles from the maser in Alexi's dilapidated Tajik house, a computer at the secret Satellite Control Facility dutifully logged the K-15's distress message and sounded an alarm. A technician ran across the room in response. He'd only heard that alarm during training, never in his thirteen years of real-world ops, though. Pulse pounding, he grabbed a red phone. "Control center here. I've got a triple-seven alert. Repeat: triple-seven alert. Platform 687 . . . yessir! A nuclear event alert."

But "nudets" or nuclear detectors carried by myriad GPS satellites in mid-Earth orbit, and Defense Support Program missile-warning spacecraft at GEO, failed to confirm that an atomic explosion had occurred in space. Minutes later, Space Control Center operators manning consoles deep inside the granite-encased Cheyenne Mountain, Colorado, complex had deciphered a flurry of signals and data exchanges. The watch commander double-checked data his operators had already triple-checked, then punched a hot-line button that connected him with senior officials scattered across the continent. One of those was in STRATCOM's operations center.

"Space Control Center update: SCF confirms EMP damage to K-15 Number 687. Pulse was not—repeat *not*—nuclear generated. Source unknown, but platform position at the time was accurately logged for troubleshooting purposes. 687's residual capability under assessment. SCC out."

12 APRIL/DUSHANBE, TAJIKISTAN

"You're sure? Absolutely sure?" Domingo's tone was threatening.

"Positive. The maser hit its target. The spy satellite has been damaged," Alexi assured the Colombian with a confident flick of his hand. Actually, the scientist had no way of knowing whether the spacecraft *had* been damaged. But he wasn't about to display a hint of doubt to this crude beast. "The tracking laser detected a variation in the platform's ephemeris data," he stated, hoping he sounded more certain than he felt. "The spy satellite is wounded."

Nadia glanced at her father, recognizing techno-bullshit when she heard it. But it had worked. Domingo hesitated, then smiled broadly and slapped Alexi on the shoulder. "Bravo, my friend! I weel tell our sponsors, yes? You prepare next shot. Ees very important!" The girl almost choked again. This bonehead Colombian had reverted to his dumbo, Latin playboy persona. Did he really believe *they* believed it? she wondered, again stealing a glance at her father as she re-configured the main control panel for a second shot. The system was still recharging. All generators had recovered nicely, she noted.

Stepping outside the modest house for a moment of privacy, Domingo searched the dark, cold sky. Nothing but stars, he half-noticed, waiting for his compact satphone to connect. *His* service was provided by a French satellite phone company, he reflected. *Can't trust those American satellites. Their calls don't go through these days!* He laughed softly, his alcohol-muddled brain finding his little joke unusually amusing.

A voice jerked him back to business. He quickly explained that the K-15 killshot had been successful, then reassured the cold, Farsi-laced voice by carefully recounting the maser-firing process, step by step. No, he didn't know the power levels used, but the Russian had assured him they were suffi-cient. Yes, you can be certain; the satellite will no longer spy on your people.

"Your peeple, they ees inveesible," Domingo assured, ex-aggerating his Colombian accent.

"The Ayatollah will be pleased," the cold voice responded flatly. Abruptly, the connection was broken.

Domingo swore softly, pocketed the phone, and plucked a cigarette from its gold-covered case. He shielded the lighter's flame with familiar ease, drawing deeply, pulling the sharp-tasting smoke into his lungs. Exhaling, he mentally deposited 5 million euros in a Swiss account. Soon, his brother would call, confirming the Iranian's electronic entry. He, Domingo, would go back inside and tell Alexi and his beautiful Nadia to fire the magic-maser again. Tonight, two American, one Japa-nese, and two Russian astronauts would die. And a second 5 million would be deposited—with an additional million on the side, just for him, in a separate, very private account. Then he could leave this armpit of a country, he thought.

Two cigarettes later, the phone vibrated quietly. Deposit confirmed. Cleared to fire.

8

REGENERATION

"Regeneration. That's STRATCOM's number-one priority," declared USAF General Erik Sawyer, addressing his AFSpace Command staff. "General Aster has tasked us, as his space and global-strike component, to restore the critical national security space capabilities we've lost. It's our job to get him what he needs, so let's hear some options."

Sawyer's suspicions were quickly confirmed, as each senior officer delivered nothing but grim news. It would take months, his staff told him, to get an Atlas V or Delta IV rocket ready to launch, but that didn't really matter. None of the new Future Imaging Architecture satellites that had been in development for years were ready anyway, so quickly replacing the K-15 Hunter in a timely manner wouldn't be easy. If another imaging bird were damaged or lost, the intelligence community would be hurting.

One spare Defense Support System missile-warning platform could be hauled out of storage to backfill the DSP lost to an earlier attack, but that didn't make much sense. The aging DSP constellation was being phased out as the new, more capable Space-Based Infrared System, or SBIRS, spacecraft were launched. Besides, the associated ground-support network was already shifting over to SBIRS. The U.S. might just have to live with a degraded missile-warning system for a while, Sawyer concluded.

But the real concern was how to restore the Navstars—the Global Positioning System navigation-satellite constellation—to its accurate, precise norm. Spare spacecraft were being prepared for launch, but rockets to fly them into orbit were still weeks from being ready. An interim capability was desperately needed, particularly in the Mideast and Central Asian regions where the U.S. still had combat forces in action. On this score, Sawyer was glad to see some serious innovation at work. A "near-space" platform was already being configured as a GPS signal-correction vehicle. It would be flown to the region within the next day or two, then readied for launch.

"Which near-space platform are we talking about?" Sawyer asked. Several were in development, following a round of prototype demonstrations a year earlier, but none was ready for fielding yet. One small company, though, working closely with a university, had come up with a novel means of getting a near-term capability into service. A staff officer quickly summarized the platform's capabilities and how it would be deployed. Sawyer nodded and said, "Great! Make it happen. This GPS fix better work, though, because we have nothing else to offer STRATCOM at the moment. Now, what about that degraded imaging situation? Any options?"

An Air Force Research Laboratory in New Mexico and the Naval Research Laboratory in Washington had offered a prototype suite of their joint, next-generation "TacSats" as a gap-filling imaging system. The scientists were confident their six-nanosatellite, distributed-imaging concept could provide desperately needed overhead intelligence information in Central Command's area of responsibility, primarily the Mideast. The system just needed a ride to low-Earth orbit. So far, a suitable launch vehicle hadn't been found in the near-term pipeline, although several small-rocket firms were scrambling to get an expendable booster ready. Nobody was sure the payload and those lifters could be mated quickly, though.

Echoing a space command chief's comments of two decades earlier, Sawyer slammed a fist onto the conference table and lamented, "Not one damned space resource is cocked and ready to go in the near term! I can't believe that we *still* don't have the space equivalent of fighters sitting on strip alert!"

Sawyer ordered his staff to keep searching for ways to reconstitute lost Department of Defense space capabilities, substituting aircraft, unmanned aerial vehicles, or other interim platforms, if those were the only alternatives.

Minutes later, he reached for the phone, dreading that tough call. He had damned little good news for the STRATCOM chief, but Aster appeared to have expected nothing more.

"Fact is, Howard, the U.S. has never gotten serious about truly responsive spacelift," Sawyer explained. "We've talked a good game for damned near thirty years, but we're still locked into a nineteen fifties paradigm. You wanna go to space? Order a rocket, then sit back and wait a year or so. Hell, I need something up there *today!* But I couldn't get a ride if I paid somebody a billion bucks. Even if I could, though, we don't have off-the-shelf payloads to replace comsats, navsats, missile-warning platforms, or weather sats on short notice," the four-star general groused. "The whole national security community depends on us, but I can put together a whole new squadron faster than I can rebuild our space infrastructure. I'm the parts man without any parts!" Aster agreed, but urged his space chief to get to the point.

"About the only good news I can offer is this: Some research-lab guys have a set of six prototype nanosats they claim can give you an interim visible and IR imaging capability from LEO. If you can spring that Blackstar spaceplane your ops guy, Forester, briefed us about, we might be able to shoehorn those nanosats into its Q-bay and get you some eyeballs in the sky. Still can't believe the spooks had that blasted thing in the barn all this time and never said one diddly word about it to us!"

"We'll work the quick-reaction spacelift issue as soon as we get a cap on this current situation," Aster said. "Right now, I need any damned thing you can get up there, 'space' or

not. I don't care how you get it, but we need comm, nav . . .
Hell, you know what we need, Buzz," Aster cut himself off,
still pacing and crushing a well-traveled route deeper into his
office's carpet. "You get those nanosats ready to roll onto a
C-130. I'll get you Blackstar. Top priority is replacing that
latest hole in our imaging capabilities." They traded a few lo-
gistical details, before Aster concluded, easing his space
component commander's lingering worry.

"Buzz, we think we have a handle on the shooter—the
SOB who's been taking down our sats and the commercial
guys' birds. We have an SOF team en route now. With luck,
the shooter will be out of business before noon your time today. So don't sweat losing everything you're now putting up.
Just get it there!" Aster signed off, then sent a mental message to that SOF team halfway around the world: *Hit that
bastard now and hit him hard!*

14 APRIL/TEHRAN, IRAN

Hassan Rafjani leaned over a small desk, smoothed to a
sheen by two hundred years of use by a long line of mullahs.
He methodically thumbed through a thin file of documents,
picked up a black-and-white photo, and carried it to the
arched window. A well-dressed, smiling Charlotte Adkins
stared back at him, a maple-leafed Canadian flag carefully
arrayed behind her.

Attractive woman . . . but in a Western way, he mused. The
shadowy Iranian that CIA agents knew only as Dagger stared
at the picture and toyed with the memory of an unusual conversation he'd had with her the day before, attempting once
again to divine hidden meaning. Arranged through the Canadian embassy, their phone conversation had revealed little.

"I would like to discuss a private offer from a very highly
placed person in the United States government," she had said.
His sharp-tongued probes had elicited nothing more, only her
skilled, diplomatic deflections and cryptic inferences to "an

offer that could ensure peace between your people and those of America."

Something in the woman's confident tone intrigued the hard-eyed Dagger, prompting him to discuss the matter with Ayatollah Ali Khamenei, Iran's supreme religious leader and the holy man Rafjani had sworn allegiance to long ago. The nation's spiritual guide had agreed that, although America could never be trusted, they probably should meet with the Canadian woman and listen to the message she carried. As a former ambassador of Canada, one of the few Western nations that maintained a sensible, understanding dialogue with Iran, she obviously had top-level contacts within the current U.S. presidential administration, and both the Ayatollah and Rafjani were simply curious. What could the new American president possibly propose that would defuse the decades-old standoff between Iran and the United States, they wondered?

Rafjani had told Adkins she would be welcomed in Tehran, although he could not guarantee an audience with Iran's president. That was acceptable, she had assured him. "I would prefer to keep this visit discreet anyway. Although I am an emissary for someone very close to the president, please understand that I am *not* representing the American government," she had said. Again, the cryptic, self-assured tone was both intriguing and amusing, coming from a brash Western woman who clearly had no sense of how a *real* woman of God should conduct herself.

The Iranian studied the photo closely. *Charlotte Adkins, you could be very useful.* He smiled. His chess player's mind was starting to formulate a plan, a strategy that could accelerate his timeline for the final confrontation, the holy battle that would fulfill God's promise. Rafjani looked forward to meeting this Canadian woman.

14 APRIL/OVER SOUTHERN TAJIKISTAN/CV-22 OSPREY

Captain Dirk Baldwin tipped his helmet-mounted night-vision goggles down to survey his surroundings—the interior of a U.S. Air Force CV-22 tilt-rotor Osprey—revealing a surreal, yellow-tinted scene when viewed through the NVGs. His small Special Operations Forces (SOF) team was strapped in along each side of the fuselage, sitting on olive-drab webbed seats. Some were dozing. Others were nervously rechecking the security of straps, pockets, and weapons. *Geared up and ready to rock,* he thought. He was damned proud of this team. *His* team, by God! He'd trained hard with these men, then led them into combat. They'd been remarkably effective at rooting out the latest infestations of al Qaeda and other terrorist groups still scattered throughout the 'Stans. Some of those ops had been dicey, but he hadn't lost a single swingin' dick on this team. *We damn well better not tonight, either,* he swore, jaws tightening.

Baldwin pulled several photos and a printout of the target from a Velcroed pocket, motioning his senior noncommissioned officer closer. With an infrared pocket flashlight, he played a flood of invisible light across the documents. Through their NVGs, which were compatible with the flashlight's IR wavelength, the two men could plainly see tonight's objective. They reviewed the planned airborne approach, where Osprey No. 1 would hover and land in a pasture, the team's one-kilometer "hump" route over a slight hill, and the route they would take to enter a nondescript residential area.

Speaking loudly to be heard above the drone of their CV-22's twin turboshaft engines and oversized rotors, Baldwin retraced the double-pronged attack sequence he and "Gunner," the NCO, had planned. Yes, they'd already been over this a half-dozen times, but you never knew when some tidbit of new insight or a previously unrecognized hurdle or whatever might crop up. Happened all the time.

Gunner recapped how he and his half of the assault team would approach the target, coordinate entry with Baldwin's group via ultrawideband covert communicators, and who was responsible for what during the takedown. The two seasoned professionals reinforced their respective understanding of the rules of engagement for using officially sanctioned "lethal force," as well as their grab-and-go objectives. They double-checked their rendezvous and pickup points and how to reach them quickly.

Then Baldwin ran through Plans B and C. Shit happened, so there were always contingency plans as backups. If Osprey No. 1 took a crap after the team's insertion, for example, Osprey No. 2, which was flying in loose formation with them now, would become No. 1 and extract the team. It would approach from a different sector to avoid a possible midair collision with No. 1. Both Osprey crews were flying lights-blacked-out aircraft and wearing NVGs that created amazing "windows in the night," but it was still damned dark out there. Man had adapted to night combat operations, yet bad stuff could happen in a heartbeat. Best to be ready. There were only so many backup resources and contingency plans available, but Baldwin felt his team had everything needed to ensure a successful mission. Still, you always wondered what you might have overlooked or failed to anticipate.

An Air Force crew chief wearing a green one-piece flight suit and flight helmet with integral NVGs picked his way through the darkened Osprey cabin, then tapped Baldwin to attract his attention. The Army SOF captain looked up and accepted a sheaf of printouts handed to him.

"These were just downlinked to us, sir," the USAF sergeant semi-shouted in Baldwin's ear. Fortunately, the Air Force's Milstar system, a constellation of secure, jam-resistant, world-wide communications satellites, was still working. The 10,000-pound birds flying 22,300 miles overhead were built to survive a nuclear holocaust. *Guess the beam-weapon bad guys haven't wiped out everything up there,* Baldwin flashed. The captain scanned a brief text message and a couple of black-and-white high-resolution images, then passed them to Gunner. The target had "remained active" following the maser shot about an hour earlier, the NCO noted.

"Whaddya make of this, sir?" Gunner asked, studying the images. One was a composite of infrared and synthetic aperture radar (SAR) data acquired by a still alive spy satellite that could "see" as well at night as during the day. Gunner had his own ideas, but the captain always read more into this sort of tech-intensive intel than he did.

"They fired that beam weapon not too long ago. Took out one of our spook sats, it appears." Baldwin pointed to a cryptic line of text. "But the site stayed hot afterward. See this fuzzy blob on the target building? That's residual heat, probably from the beam-gun or other equipment in that greenhouse on the roof. Some of our big-brain intel dudes think the bad guys're gonna fire that sumbitch again," Baldwin said, interpreting the text's somewhat convoluted intelese. "Way I see it, Centcom's basically telling us to get our asses in there ASAP and whack these beam-freaks before they zap something else."

Gunner nodded. *What the hell do you think we're doing, assholes? This taxi's gettin' us there 'bout as fast as it can.* The battle-hardened NCO was, once again, amazed at how headquarters always felt obligated to restate the friggin' obvious, the physics of time and space be damned.

The USAF crew chief was back, leaning toward Baldwin's ear again. "Twenty out, sir. We'll start our descent any minute now. Major wants to see you, though."

Baldwin swore, but staggered to his feet, ungainly with all the gear his thick, muscular form was carrying. He left a rucksack behind, but gripped a short-stock, modified M-4 rifle. If this bird crashed, dumping his team into the middle of a firefight, he sure as hell wanted the weapon handy. Seconds later, he stuck a helmeted head into the pilot's flight compartment. Instruments barely glowed on the panel facing two pilots. All flight-deck crewmembers were wearing NVGs. A navigator waved Baldwin over, turned on an IR lamp and pointed at a photo-map combination. Another computer printout, Baldwin noticed.

"Got this update when your stuff came down, Dirk. Everything looks good for the insertion, but, if we have to come get you, it looks like the backup extraction point has some new obstacles. Might make it tough getting in there," the navigator said. "The boss says he'll try, if you absolutely need it, but we'd rather set down over here." He pointed to a nearby opening among the jumbled set of houses. "That'll probably wake up more of the natives, so we don't want to jump in there until your guys are on site, ready to board. That work for you?" Baldwin nodded, rehashed the details one more time to ensure everybody was on the same page, then headed back to the dark cabin, picking his way through equipment and outstretched legs.

He quickly outlined the revised secondary pickup point for Gunner, then huddled with his own assault subteam while Gunner did the same. "Questions? Everybody got it?" Baldwin finally asked. Whites of eyes set in camouflage grease-smeared faces bobbed up and down in the near-dark. "Okay. We'll be on the ground in . . . eleven minutes," he said, checking his wristwatch. "Make sure you activate your comm, then check in as soon as we're outside. This is big stuff, guys, so let's do it right."

Baldwin stood and slapped the rock-hard shoulder of a stocky soldier before returning to his own position. He scanned the aft end of the Osprey's cabin, noting the open space allocated for whatever they might grab at the target and haul home. He took a deep breath to calm himself and wondered again: *Exactly what the hell are we going to find down there?*

14 APRIL/DUSHANBE, TAJIKISTAN

Alexi shivered involuntarily, but not from a chill that permeated the weapon control room. Domingo's hard-eyed, no-nonsense demeanor had returned, and the Colombian was demanding the second maser shot be taken immediately.

"But Mr. Domingo . . . the target is *not* visible in the sky right now. There is nothing for us to find, track, or lock on to!" Alexi tried to explain again. Domingo was red-faced, threatening to explode, when Nadia stepped forward, holding a ball of Styrofoam packing.

"Look. The target is here; we're over here," she explained

quickly, holding the ball in one hand and pointing with the other. "We can't fire until the station is over here. Otherwise, we'd have to shoot through the Earth. And that's not possible." Very simple, but graphic.

A light flashed briefly in Domingo's cold eyes. He was silent for several seconds, staring, before his face darkened again. "How long do we wait? Until you *can* shoot?" he demanded. The mindless, playboy persona had long disappeared, replaced by a brutal, narcothug attitude.

Alexi had already pulled the American space catalog up on his screen, and had the International Space Station's position highlighted on an Earth-orbit graphic. "The . . . uh . . . objective . . . is in a 250-mile-high orbit, Mr. Domingo. It's also inclined 52 degrees, so it makes this oscillatory track north and south of the equator. It will fly over us in several hours, but—"

"We cannot wait several hours, you idiot! The Americans could be here any minute! You *will* shoot . . . ," Domingo screamed, pounding his fist against the tiny room's wall.

Alexi interrupted, irritably pointing to the screen. "Yes, yes! As soon as it clears the horizon enough, we will fire. That will be in . . . twenty-two minutes. Either we wait until then, or we fire for nothing. You see?" The Russian scientist sounded much calmer and more certain than he felt, but his confident self-composure had the intended effect. Domingo, though still agitated, now seemed resigned to the facts of orbital mechanics.

He finally flicked a hand in disgust. "Okay! But you will fire as soon as possible! There is very little time!" He slumped against the wall, checking his Rolex Oyster.

14 APRIL/LANDING ZONE/DUSHANBE, TAJIKISTAN

The CV-22 descended steeply, nose up, its wingtip engines and huge propellers swiveling from horizontal to vertical, into helicopter mode, repositioning the rotors overhead. Seconds later, the awkward-looking machine settled onto the spongy grass of a rolling pasture, the rear ramp dropped, and a dozen shadows disembarked, crouching and on the run. Baldwin's SOF team scattered. Several men quickly formed a perimeter, covering their compatriots' exit. A signal from the young captain and the Osprey lifted, turned, and joined its backup aircraft, which had continued to orbit a mile to the south.

Baldwin and Gunner quickly split the team into two predesignated groups and left at a trot, topping a nearby hill only after two advance scouts quietly radioed an all-clear. Moving with surprising speed, the ghostlike figures flowed around silent, darkened residences and converged on the building intelligence agents in the U.S. had identified some forty hours earlier. Using hand signals visible only to those wearing NVGs, Baldwin silently directed several men toward the rear of the building. They'd cover any escape routes while the primary team went through the front door. Via an earpiece and tiny boom microphone, he and Gunner whisper-coordinated the prepositioning of every team member.

14 APRIL/INSIDE THE MASER SITE'S BUILDING

Time moved with agonizing slowness beneath Domingo's glare. But finally his target's icon appeared on the screen.

Alexi tracked, waited a few seconds, then locked on to the International Space Station as it rose above the southwestern horizon. *A few more degrees of elevation . . .* He was holding his breath, waiting. Nadia double-checked her equipment and maser system readouts before announcing: "Capacitor banks fully charged; power at maximum. Ready to fire." Domingo hovered behind Alexi, trying to will the small ISS icon on the Russian's screen to climb faster.

Suddenly, the Colombian's cell phone rang, startling all three. Domingo stepped to the door, his back to Alexi, before answering. Then he turned and screamed, "They are here! *Fire!* Fire now!" A Tajik liaison officer aboard one of the now orbiting USAF Ospreys had radioed his headquarters, telling his commander that Baldwin's SOF team was on the ground. Long ago, Domingo had cultivated a low-ranking Tajik communications system operator in that headquarters, paying him well to ensure the civilian would relay valuable military and police information on a moment's notice. That info had helped avoid embarrassing situations with certain women and underworld figures before. But tonight, the piddling euros paid to that comm operator had bought the Colombian a few minutes of critical warning.

Alexi turned to Nadia, who was poised, thumb under the switch. He swallowed hard, setting a prominent Adam's apple into motion, and ordered, *"Fire!"* Nadia jammed the switch upward, then spun at the sound of a loud crash. Domingo twisted toward the noise, reaching inside his coat as he bolted for the hallway.

Two of Baldwin's SOF team kicked in the house's front door, then dived into a darkened room, rolling to a kneeling position on each side of the door. They whispered "Clear! Advance!" into their communicators' boom-microphones. Baldwin and three other dark-faced soldiers were instantly inside, leapfrogging the first two's position and flattening their bodies along the living room's wall, opposite the outside-entry doorway. A yelp was followed by a sharp report from Domingo's 9-mm Sig Sauer 26. A doorframe opposite Baldwin's face splattered wood splinters into the room. Then all hell broke loose: bodies rolling through the door, fire spitting from short-barreled automatic rifles, shouted commands, and a piercing scream. Then it was quiet. The heavy, acrid odor of spent ammo hung in the air.

"Alpha's secure. One enemy shooter down; wounded. No casualties. Under control," Baldwin radioed. "Moving deeper . . ."

Gunner answered immediately. "Perimeter secure. No response from the natives. Holding position."

Inside the control room, Nadia had grabbed Alexi's arm as soon as Domingo pulled the automatic pistol from his shoulder holster. Thinking her father was the Colombian's target, she'd instinctively screamed and jerked Alexi to the floor. She threw her small, thin body on top of her father. Head buried, she'd flinched every time a shot was fired, but refused to let Alexi get up, shouting in Russian "Stay down!" When the shooting stopped, she scrambled beneath a workbench, trying to pull him with her. She suddenly froze as a black-faced, helmeted figure dived through the room's only entryway, dropped to a knee, and pasted itself against an equipment rack.

Nadia's pale blue eyes were huge, locked on to those of the invader. She slowly raised her hands, palms out, silently pleading for her life. A small, round hole in the soldier's rifle barrel never wavered. It remained focused on the girl's forehead, tracking her every move. Unbeknownst to her, a tiny red spot of laser light quivered on her forehead. Then another voice, firm but quiet, reached her. "Do you speak English?"

"Yes! Yes! I speak English!" she squeaked, trying to swallow the abject fear that consumed her. Alexi sat up, but cracked his head on the bench. Nadia spoke sharply, in English. "Father! Stay down! Do not move!"

Baldwin was crouched in the doorway now, his M-4 held at the ready. "Both of you! Face down on the floor; hands behind your heads! *Now!*" The other soldier still knelt by the rack, his weapon also covering the two Russians.

Soon, Nadia and Alexi sat against the small room's wall, their hands bound behind their backs by plastic wire-ties. They both turned as Alexi's wife appeared, herded in by another soldier. Wrapped in a thin housecoat, hair askew, the woman was scared speechless, barely able to walk on her own. She uttered an exclamation and dropped beside her husband, chattering and crying in Russian as she tried to melt into him.

Baldwin felt a twinge of sympathy, but set it aside. He motioned for the soldier to also bind the older woman's hands, then turned to a trooper huddled over a computer screen. "Get that thing shut down and unhooked. It's going with us," he snapped.

"Sir, you'd better look at this first," the soldier said, pointing.

"Looks like these shitbirds got a shot off before we hit the door."

Baldwin scanned the graphic, then glanced at a block of text in the screen's corner. "Oh, son of a . . . ! Is that—?"

"'Fraid so, sir. I think they just zapped the space station."

9

CASUALTIES

They never had a chance.

When the powerful maser pulse struck the huge, gangly space platform drifting above Central Asia, only one astronaut was awake. The station's lights dimmed, then extinguished. Instantly, faint emergency lights popped on, providing illumination the astronaut needed to start a rapid check of the electrical system. The station was already on reserve power. He confirmed the main power controller was inoperative, the equivalent of massive circuit breakers thrown to the OFF position. He hesitated a few long moments, reflecting. He'd never seen a power failure this extensive. The panel before him indicated the entire station had gone dark.

Flashlight in hand, he grasped one of several U-shaped handles on a bulkhead and yanked, launching his near-weightless shape through the main bay. He drifted into the sleep area and noticed four forms trying to free themselves from cocoon-like bags that held sleeping astronauts in place.

"What's the problem?" the American station commander asked, finally floating away from a formless bag hovering in its compartment. He wore a blue zip-up sweatshirt and matching shorts, as well as white athletic socks.

"Some kind of electrical spike. It appears we've lost the entire power system," the Japanese crewmember answered. "Very widespread. We're on backup battery."

"The environmental system. Is it still operating?" asked a burly Russian, the only ISS veteran in the four-man crew.

"I did not check. I thought I should get all of you first. This could take time . . ."

"Good move, 'Ink,' " the American assured him. He issued orders for others to check various systems, then propelled himself toward the main work bay.

The five astronauts quickly assessed the station's various systems and compared notes. The electrical power-generation network was "hard-broke" as the American put it. All but one communications package, a backup unit that had been unpowered when the pulse struck, were dead. The environmental system was definitely damaged, but to what extent it was difficult to tell without further diagnosis. They'd reported the station's status to a ground station in Russia, and were now patched through to a rapidly growing cadre of engineers, managers, and specialists in Houston and Moscow.

Finally, the ISS program manager in Texas issued an order every astronaut crew had prayed they'd never hear: "Abandon the station. Get in the escape vehicle and execute the return-to-Earth sequence. Do *not* delay!"

The astronauts quickly gathered a few critical safety and personal items, then crowded into a cramped Crew Escape Vehicle docked to the side of their spider-like station. Sealed inside, systems activated and cleared to disengage, the American commander threw the correct switches. Nothing. No comforting "thump" that would have signaled their CEV had pushed away from the wounded ISS. After a quick review of the disengage procedure, a second attempt. Again nothing. Pulses racing, the crew tried the manual release process, but the CEV refused to budge. It was locked to the station's docking port.

In Houston, the ISS program manager scanned several computer displays, then straightened and turned to his deputy. "If we can't get that damned CEV loose . . ." The deputy stared back, then slowly shook his head.

14 APRIL/CREW ESCAPE VEHICLE/INTERNATIONAL SPACE STATION

Oxygen continued to course through the CEV, and a few dim lights on a control panel remained illuminated. But there was no longer a hint of movement inside. The crew members were motionless as the capsule's temperature rapidly dropped. Without power to keep heaters running, and unable to draw electrical sustenance from the dying station, the CEV's environmental control system slowly failed.

At first, the men had shivered, their teeth chattering uncontrollably. Then they grew tired and lethargic, the deadly symptoms, they all knew, of freezing to death. They had been trained to fight to the end, to always make the impossible happen. But every man quietly accepted the truth. Like the space station and the CEV firmly locked to it, the men were dying.

They sent a few final, somber words to their families via Houston and Moscow—good-byes, blessings for wives and children, last wishes, and, finally, prayers. Then each succumbed, drifting off to sleep. Within an hour, all five were frozen solid, victims of space's hostile, unforgiving cold. It would be a month before a rescue vehicle—the only remaining space shuttle still in service—could be launched from Cape Canaveral. But it would no longer be a rescue mission.

14 APRIL/STRATCOM HEADQUARTERS

The astronauts' final seconds of life had already passed when General Howard Aster took the notification call, then slowly cradled his phone. He stared out the large window, stunned, yet uttering a silent prayer for the lives that had been taken, before speaking softly to Burner Burns, his aide. The latter knew it was bad news, but hadn't been able to decipher the message from his general's one-sided conversation.

"That sumbitchin' maser hit the space station. They couldn't get the astronauts off," Aster said, his voice barely audible. "They all froze to death, Burner."

Burns was speechless. *The space station? How . . . ? No! No way! That can't happen!* He had an uncontrollable urge to hit the RESET button; to go back and replay the last few seconds, to set it right somehow. Surely he hadn't heard correctly. Then full realization struck. *We really are at war. Space war! . . . But who's the damned enemy?*

Heroes had been murdered. Somebody would pay. The American public would demand blood now. "Space war" was no longer a string of mere day-to-day inconveniences. Astronauts had died before, but those were accidents, tragedies attributable to the risks of exploration. But not this time. These pioneers from several nations had died as victims of a cold-blooded, premeditated attack. Private citizens in hundreds of different countries, who, a few hours earlier, couldn't have told someone how many people were even on the ISS, would now see those five astronauts as heroes whose deaths must be avenged. International demands for retribution were guaranteed.

Burns noticed Aster glaring at him. "A threshold has been crossed. Get the battle staff and 'gamers together in the center, Burner. We're going to a new level. First, we'll wrap up this DEADSATS game, then move on to the next issues. This isn't over yet. . . ."

14 APRIL/OSPREY NO. 1/AIRBORNE OVER TAJIKISTAN

Baldwin, still on a high from his team's successful strike and takedown, now stared at the printout a crew chief had handed him, disbelieving. *You gotta be shittin' me!* Even forewarned, before they'd departed the maser site, he couldn't fathom what he was reading. But CENTCOM wasn't in the habit of playing twisted jokes, especially via Milstar comm links to

CV-22s winging their way home after a very successful "black-ops" mission. As full understanding sunk in, Baldwin slowly leaned back against the Osprey's bulkhead, suddenly fatigued beyond human measure. *We were too late!* He quickly scanned the scrap of paper again, mentally converting Zulu or Greenwich Mean Time and comparing the figures to local time. *Aw shit! Shit! We were there; right there! A matter of seconds! Only seconds, for chrisakes!*

The young captain groaned and closed his eyes, but only saw five faceless, frozen, blue-suited astronauts swimming before him. *I killed them.* The thought from nowhere was crystal clear. Not shaded with doubt. No twinge of uncertainty. *If I had given the "go" order just . . . two minutes earlier? They'd still be alive. That beam would never have been fired. I, Dirk Baldwin, killed them!* He was overwhelmed with bottomless guilt and a soul-deep sense of failure. In a matter of minutes, his being had swung from the adrenaline high of savoring a perfectly executed, incredibly successful mission to a grief and self-recrimination he'd never experienced before. He had failed those innocent men 250 miles above. If only he'd acted sooner!

Baldwin opened his eyes, startled to see Gunner kneeling directly in front of him. "Sir, what the hell's wrong? You look like someone gut-shot ya!" The captain stared at him, befuddled mind toying with a ridiculous thought. *How can Gunner see me in the dark? Oh. Goggles . . .*

Shaking his head, trying to clear its brain, Baldwin held the printout for Gunner to read. The NCO scanned it, then registered the same shocked realization. Quick comparisons confirmed each man's gnawing fear. Although right on their preplanned timeline, they'd been only minutes, maybe *seconds,* short of preventing the ISS tragedy. They were both silent for a long few minutes. Then Gunner slowly looked sideways at his commander, jaw clenched.

"Sir, I think I'll have a word with our prisoners," the NCO said, rising. Baldwin instantly connected and stood, as well. The two men picked their way back to where the three Russians were huddled together. Baldwin had ordered the three kept together, under guard. He'd briefly interrogated them at the target house, conducted while the bulk of his team packed key computers, files, and specific equipment that Nadia had pointed out as important components.

Domingo lay on the opposite side of the aircraft's cargo bay, his hands bound together. He'd taken a round in the right shoulder, but a medic had patched him up and jabbed a needle in his leg to ease any pain. The Colombian had tried to reach a hidden knife before his hands had been bound, but an alert soldier had stomped on the drug lord's wrist, crushing the scaphoid and lunate bones into fragments. The knife quickly disappeared, and Domingo now nursed a perforated shoulder *and* a smashed wrist, both still throbbing with pain. He glared at the Russians, his mouth twisted in a sneer of hate. The cargo hold was darkened, but chemical light-sticks taped to soldiers' legs and structural frames near both the Russians' and Domingo's positions cast a ghostly green glow.

Baldwin knelt and spoke quietly, first to Alexi, then Nadia. The girl answered without hesitation, shaking her head vigorously, side to side, in response to questions. Finally, she gestured toward Domingo, rapidly explaining something. Baldwin nodded, then stood. He and Gunner talked briefly, while both eyeballed the Colombian prisoner. They approached him, faces grim. Domingo felt raw fear rise in his throat. Something in the grizzled NCO's eyes told him Special Ops teams didn't always play by Geneva Convention rules. Domingo, who had reveled in the fear of others, knew it was now his turn.

Baldwin asked the narco-killer several questions, but Domingo reverted to his playboy Latin mode, whining and shrugging repeatedly. "I do not know, meester! I just de mesng'r, ya know? I know notheeng!"

Baldwin stared hard at the fat man lying on an Army stretcher, one shoulder and a wrist bandaged, but covered with a fine-quality, expensive suit jacket. The captain tilted his chin toward Gunner. A barked order from the NCO brought two burly soldiers into view. One on each side, they lifted Domingo bodily and carried him aft. Gunner opened a side parachute door, admitting a blast of cold, violent air. Domingo shuddered, eyes growing wide with terror. The soldiers, each holding Domingo by a leg and his coat's lapels, placed the man on the deck and edged his head into the open doorway. Vicious, howling wind clawed at the prisoner's thick hair, trying to rip it loose.

Domingo screamed a string of Spanish, eyes now bulging in fear. Though bound, his one good hand tore at a soldier's arms, the same powerful arms that pinned him to the cold metal cargo deck. Baldwin knelt close, held an index finger before Domingo's nose and yelled. "*Otra vez, gordo.* Who paid you to hit the space station? *Y porqué?* One chance at the right answers. *Uno! Comprende?* Or you take your first flying lesson. *Ocho bolard!*" he added for emphasis.

Baldwin held a knife-edge palm aloft, ready to drop it if Domingo failed the simple test. The Colombian promptly wet himself, bringing a curse from the downwind soldier. The other nudged Domingo's hulk another inch into the open doorway, letting the 230-knot wind buffet the man's cranium even more. Domingo screamed a long litany, still in Spanish. Baldwin glanced across Domingo's body, and caught the quick nod from Sergeant Perez. He'd understood. Baldwin turned back and yelled again.

"*Perdón! Mi español poquito. Inglés ocho bolard!*" With a glance and tip of his head, Baldwin motioned to the terrifying blackness below.

Domingo responded immediately, this time in English, spilling every tidbit he ever knew about anything: his Iranian contact's name, the Swiss account, drug deals in the United States, his side deals with the Iranians, why he wanted euros instead of dollars, and which spacecraft the maser had targeted. Baldwin grabbed a handful of the man's coat and yanked him upright, but still close to the open, yawning door. Making sure Domingo could hear clearly, the young captain shouted several more questions about those Iranian contacts and Swiss bank accounts, and Domingo's brother in Colombia. Finally, he signaled Gunner to close the jump door.

"*Gordo,* you'd better be telling the truth. We're going to relay this to our people. They'll check it out. If you're bullshitting me—and I'll know before we land—I'll personally throw your ass out. *Comprende?*" Domingo nodded vigorously, thanking Baldwin profusely. He was sweating madly, despite the cold. The stench of fresh urine quickly spread through the cargo area.

Gunner had his soldiers lash the Colombian to a tiedown ring near the door. "We don't want to lose our primary witness, right? He might just decide to fly on his own. But I doubt it. What a wuss," Gunner snorted, turning to follow Baldwin.

14 APRIL/THE PENTAGON/SECRETARY OF DEFENSE OFFICE

With a telephone receiver clamped to one ear, T. J. Hurlburt grunted, nodding now and then. He slumped over the expansive desk, one hand rubbing his forehead. "You're absolutely sure this . . . this drug creep is on the level, Howard?" He listened intently to his STRATCOM chief, nodding, then smiled and chuckled softly. "Okay! I get the picture. Tell that SOF captain the SecDef can *not* officially condone such extraordinary measures. But this old soldier sure as hell does! My congratulations to that boy and his entire team.

"There's no way they could have known about the station being targeted, so I want those kids to get their damned chins off the floor, too!" he growled. "They pulled off one hell of a mission, and I'm damned proud of 'em!" He slammed the phone down and leaned back, reflecting.

Hurlburt hooked one heel on an open desk drawer, grabbed the phone again and started rattling off orders to his deputy. "The spooks are interrogating the Russians and that Domingo weasel now. Make sure everything they squeeze out of those bastards gets distributed to the right folks, especially the DEA and the FBI task force. I want a summary of the same info. And sic the lab rats onto those computers, too. I want every friggin' digi-byte in those Russian machines wrung out and scrubbed by an interagency intel team. Better put some tech-intel types on that other gear the SOF guys hauled back, too. Whatever they learn about that maser, run it by Aster's space-control folks at STRATCOM. Howard will be very interested in the Russians' sumbitchin' maser. It's still hard to believe such a peewad outfit could cause so much friggin' damage to our space assets!"

Hurlburt hung up, then rubbed his face with both hands. *God, what a mess! How the hell did we get into this? Two smart Russians, a pea-brained narco-terrorist, and a Tinker-toy homebuilt maser. And they brought the most powerful nation on Earth to its space-knees! Even Hollywood couldn't dream up such an outlandish scenario!*

He glanced at his watch, swept a few notes into a folder, and headed for the door. The president was *not* going to appreciate his morning security briefing. The maser and its immediate threat were out of business, but nation-state progenitors also had a hand in this mess, and were still out there. Five astronauts were dead, and the space station was a derelict drifting on the black, airless seas far above Washington. That all led to a tough question: How the hell should the U.S. respond?

15 APRIL/STRATCOM WARGAMING CENTER

"Okay, ladies and gentlemen, this is what we in the wargame business call the 'hotwash' session," Androsin said, addressing an auditorium filled with uniforms and suits. "You're all aware of the ISS tragedy, and you know the maser site has been neutralized. But we're not finished. We still need to go through the wargame wrap-up, because there's a lot of lessons learned here. The 'hotwash' is where we recap the game's precepts, its findings, and team recommendations as to where we go next."

He laser-pointed to a list displayed on the center's main screen. "We'll summarize what your groups came up with, attempting to glean some insights that will help us and the battle staff in our next phase. We'll bounce that against information from the SOF team, the intel folks who debriefed the maser crazies and anything else BOYD thinks is relevant.

"For those of you who have never participated in a wargame before, let me remind you that there are no right or wrong outcomes," the colonel cautioned. "Our ultimate intent is to develop a range of options for decision makers. We first want to gain insights and understanding of the issues that we've explored, though. Keep in mind that we *are* at war. Even as we speak, our degraded space situational awareness and our crippled GPS network are affecting the nation's entire geopolitical situation."

Androsin handed the wireless microphone to Colonel Matt Dillon, who retraced the game's process, pointing to each bullet on the computer-generated display. The game had taken a few unanticipated turns, but, overall, had stayed on track.

General Howard Aster sat in the front row, concentrating on everything each speaker was saying, occasionally jotting a few notes on a small pad. Tanner was now at the podium, gesturing with a laser pointer.

"The real breakthrough was when the Attorney General's Office told us about a joint investigation under way by the Homeland Security, Justice, and Treasury Departments—DEA and Customs. Seems the DEA office down in San Diego was running a money-laundering sting on Colombian cartels that had set up bona fide businesses in the U.S. That's the typical way they conceal and clean up drug money. Customs got involved because money was moving around the world and back into the States, only it'd been converted to euros by then. Seems simple, but it's not. And an Israeli consultant to the task force figured out the activities of a Middle Eastern family banking operation. I asked Agent Frank Donovan from Justice to run us through how it worked."

Donovan blew into the mike a few times to make sure it was working. Although impressed by the high-ranking military officers in the wargame center, he was hardly intimidated by them. He explained why the space-insurance scam seemed to make no sense at all—until Beatty showed how the numbers had ramped up from satellite losses.

"You see," he said, "money laundering is supposed to *protect* the cartel's cash, not *lose* it. So, at first, it just looked like plain old bad business. But this company wasn't dealing in lost dollars. It was ramping up rates, then converting its currency into euros, as if they knew something about the sinking value of the dollar. This company, like most fronts, looked and acted legitimate—so legitimate that it had become a major underwriter of space ventures. They'd taken aggressive positions on several commercial satellites, and had just started writing policies for some of the new reusable-launcher start-ups. Their staggering profits had actually spurred competition with other, more established insurance firms, who decided they'd better jump into space.

"And these bad guys were slick. Their business acumen was what threw us off track, at first. So we looked to the space insurance companies for help. Mr. Beatty can go over some of the details of how that operation worked," Donovan concluded.

Merle Beatty, the wargame's space insurance industry representative, slowly climbed the few small steps. He appeared tired and older than his age. When he spoke, though, years of experience were apparent. "This cartel-operated company, Space Underwriters, Ltd.—which is now the target of a federal investigation, by the way—was *not* a member of our insurance association. An important clue. Once we saw that they were incurring big losses from these satellite failures, and weren't part of the industry's self-protection mechanisms, we started comparing notes with the feds." Beatty glanced at Donovan, then turned to the big screen, now displaying a mass of linked ovals.

"It looks complicated here, but is really surprisingly simple and clever, I must say. The key is taking the long view, something insurance people always do but investment bankers rarely do. Once we sat down with the investment guys, that federal task force and commercial space companies participating in this wargame, we figured out what was going on. No single entity had the whole picture, but, collectively, we were able to piece it together. Especially when Agent Donovan explained the money-laundering angle," Beatty continued.

"Take a look at this chart. And remember the whole point of money laundering. The bad guys take drug money that they can't get back into Colombia, and comingle the funds with a legitimate business they invest in, so the money looks like it's generated by above-board business transactions. By the time they pay their taxes, there's no difference between dirty money and clean money. Case in point is Space Under-writers, set up entirely with drug money and, ultimately we came to find out, with Iranian financing. They paid out millions for failed and limping satellites. That gave them a good reason to raise rates even faster, and demanded *higher* rates than the rest of the industry.

"They were already committed to insuring more than fifteen upcoming launches of spacecraft that used the Boeing HS-601-series satellite bus. So a few-percent increase on each one yielded a nice little profit. And it looked like they were taking quite a risk by loyally sticking with satellite companies at a time when nobody else wanted to issue a policy on *any* -601-based bird. They looked like good guys, while knowing full well that there really wasn't any risk involved.

"They were playing both sides and could order which satellites to take down!" He bristled. "They took out a few government birds, too, as a way to hide their real objectives. *They* were the ones causing the satellite failures! They kept issuing new policies, using illegal drug money as capital to pay off the failed birds' owners, while reaping a super return on higher premiums. Imagine: laundering illegal money and turning a big profit in the process! These guys were set to make billions in untraceable dollars, which they then converted into euros to satisfy some Iranians—who the feds are still trying to identify. Their Iranian connection flowed counterfeit U.S. currency into the international markets. Agent Donovan eventually figured out that we were taking a double hit: the Iranians were weakening the U.S. dollar, while their Colombian partners were taking down our communications, surveillance, and global positioning satellites that, we believe, blinded us to what the Iranians were doing. But I'll leave that to the rest of you."

Aster stood, signaling it was his turn on-deck. His long stride took the stage two steps at a time. Eight stars, four on each shoulder of the crisp, light-blue, long-sleeved shirt, sparkled under the platform's bright lights.

"Obviously, I'm no rocket scientist, but I wanted to do the wrap-up—even go over some technical stuff—on what was killing these birds. And I need to put a few things in perspective," the general said, pacing across the stage. "Thanks to our intel friends and ILM, that environmental outfit, we narrowed down the source of these killer signals to one building in Dushanbe. In parallel, the DOJ and Homeland Security task force and the CIA were still working with George's team, trying to figure out the insurance connection you just heard about. Thanks to some skillful negotiations by State, the Tajikistan government actually had its military and police forces help us pin down the maser's location.

"But then we faced a dilemma: In whose jurisdiction were the perpetrators? Was this a Tajik issue, or was the U.S. justified in launching a covert military mission to defend its space infrastructure? Had a crime been committed, and, if so, how could the perpetrators be prosecuted? Under what laws, by whom, and in what court? But that came later. The first order of business was figuring out what the hell the perps were doing.

"Turns out that the bad guys' maser—sort of a microwave laser that shoots a high-powered beam of electromagnetic energy—was a hell of a lot more effective than the particle beam weapons we toyed with during the 'Star Wars' days," Aster continued. "Now, this maser beam doesn't always travel in a straight line, like a laser does, so hitting a satellite at geosync orbit, 22,300 miles in space, with enough energy to have a significant effect, isn't trivial. The drug cartel and their Iranian backers spent a lot of bucks supporting a Russian ex-scientist and engineer they'd recruited. They were trying to perfect a high-tech weapon system that the Soviets had been working on back before the Berlin Wall came down, and it looks like they did a fair job of getting there," the general explained, still pacing with a long, slow stride.

"The cartel sort of dared the scientist to take down some of our GPS birds, then the DSP missile-warning platform in GEO. That, folks, was *not* easy and should scare the hell out of us. All of our 'expert' scientists told us a maser couldn't be controlled that precisely, *and* couldn't possibly reach GEO *and* absolutely could *not* damage our super-hardened milsats. But it sure as hell did.

"The bottom line is this: The bad guys were able to induce abnormally high electrical currents in all targeted spacecraft, which started a chain reaction of internal system failures. You've already been through that," the general clipped. He aimed a laser pointer at the screen and ticked off impacts the disruptions had caused, setting up his audience.

He asked that the display be turned off and slowly walked to the center of the platform, head down, hands in his dark-blue pants' pockets. "Finally, the bastards managed to hit the space station, triggering an international disaster that exceeded NASA's and STRATCOM's worst space nightmares. No threat analysis, no systems-engineering analysis, *nothing* had anticipated the effects of a massive electromagnetic pulse on the station. You'd think that would have been done, given the threat of a nuclear detonation in space, but . . . Hell, that's a technical issue." He dismissed the subject with a hand flick.

"Now, folks, let's consider what happened here. A non-government-sponsored entity damaged, and in some cases disabled completely, almost $2 billion worth of commercial satellites. Not all were lost, of course, but enough disruption was caused to severely impact a half-dozen or more spacecraft operators. Throw in the stock-price hits on several huge companies, loss of service to customers, higher insurance premiums, delayed launches or service-starts and other factors, and you have several million more in costs. Add a few billion for the government DSP, GPS, and national technical means birds, and, of course, the International Space Station, and you have a major, major impact on the U.S. and international economies *and* space infrastructure. All caused by a few narcoterrorists doing what narco types have been doing for the past thirty-five years: laundering money. Only these guys were thinking well outside their usual sleazy box."

The general stopped, legs spread slightly, hands on hips as he faced the darkened auditorium. He was transformed into the classic fighter-jock commander, addressing his flock after a successful mission. "How did we solve the problem? Well, we lucked out, plain and simple. Our wargaming tools proved to be fantastic aids, especially when we fed the latest intel to you, the smart players. But we were still damned lucky."

Aster paused, then resumed pacing, before continuing in a softer tone. "Unfortunately, the damage extends much further than we could have dreamed. Now, a number of state actors are taking advantage of our wounded space posture. On the geopolitical scene, a lot of nasty crap is surfacing, simply because some rogues smell U.S. blood. Like sharks, they know we're wounded, and are striking while we're limping—in a space sense. None of us here at STRATCOM, or even at the smart-guy think tanks in Washington, had an appreciation for just how powerful a global deterrent our space resources have become. Now, some of those bad actors are causing one hell of a lot of mischief, and we're going to have a devil of a time getting a handle on the situation. Our leadership is talking about 'tipping points' and 'spinning out of control.' All because we lost some very important space platforms.

"But the real eye-opener is this: We've been waging war against an entirely new enemy, one that we had failed to identify or anticipate. In a way, it's 9/11 all over again. This enemy wasn't a front-line nation-state. These guys were a criminal cartel armed with tons of illegal money. Yes, they were in cahoots with Iran, which is still a serious threat, but the crooks could have pulled it off on their own. Nowadays, that's all a potential adversary needs, because he can buy technological expertise on the black market. Consequently, the drug cartel was able to acquire sophisticated weapons and outstanding for-hire expertise. That small bunch was systematically taking out U.S. economic and national security assets, and we were damned near powerless to do anything about it," the general continued.

"Now, how prepared were we to deal with this situation? Not very well, I'm afraid." Aster started snapping his right forefinger against successive digits on the other hand. "One. STRAT-COM had no sure way to even verify that our satellites were being interfered with. We had no proof to give our own command authorities, and none to back up our bitches on the international political scene. We *gotta* have new sensors on *all* these birds, people! We're working on that with the Pentagon and trying to convince Congress to fund an even more robust suite of attack-alert sensors on every new military and civil satellite we orbit in the next few years. And you commercial guys need to work with us. Let's find a financially viable way to get these on your spacecraft, too.

"Two, it took a multi-agency government task force, plus a lot of you commercial-sector experts, to identify the fact that a problem even existed, the source of that problem, and what to do about it. Nobody could have done it alone." Aster was pacing again.

"Three. We had no established policies to guide a response, once we had nailed down the cause and the source of those attacks. And still don't, frankly. We quickly learned that the military had no charter or clear process to take action. We couldn't protect our on-orbit assets, and weren't sure we had the rights to launch a strike against the bad guys. Our CIA, DOJ, and Homeland Security guys couldn't do anything, either. At least nothing legal." A ripple of quiet laughter sprinkled across the auditorium.

"The commercial companies whose satellites were being targeted needed protection for their expensive assets, but were reduced to waiting for someone else to take action against the group firing that maser. Our Drug Enforcement Administration and others were already going after that particular drug cartel's money, but having little to show for their efforts. Ultimately, we resorted to political pressure on the international scene, plus the SOF takedown mission, pretty much making things up as we went. That put the cartel and its scientists out of business.

"Okay, that's the good news." Stern-faced, Aster paused, then bellowed, "Folks, I'm here to tell you, we're now in very real danger of being dragged into a shooting war! All because we lost vitally important space-based assets!" The general paused, eyes flashing as they took in the dead-silent crowd.

"STRATCOM absolutely must have more options to exercise space control. We need national and international policies, doctrines, and . . . mechanisms, let's say, that allow us to quickly protect any U.S. or allied satellite that's under attack, then respond to those attacks. We have some classified means on the books now, but they're pretty much terrestrial options—special forces operations against a bad guy's ground sites; jamming of the interfering signals; diplomatic pressure; and a few that I can't talk about. But the day's coming when we'll need to fly a space fighter up there and literally defend a constellation of satellites, an outpost on the Moon, or some other asset from threat missiles, masers, lasers, or God knows what. Maybe even fight another space vehicle, like a killer satellite that tries to ram or blow up one of ours."

Aster grinned slightly and his voice softened. "Yes, that sounds pretty far out, but I can tell you, we're headed in that direction. Jack, you have something to say?" the general interrupted himself, motioning to TransAmSat's vice president, who had a hand in the air.

Molinero stood laboriously. The overweight man always looked weary, Aster reflected. "General, I just want to reinforce a point and ask a question. I second the need for putting attack-alert sensors on every commercial satellite that goes up. Yes, it'll cost all of us, in dollars and weight and power, but we have to do it. In fact, it makes financial sense to do it as soon as we can. If we save one or two satellites in one company's fleet just by knowing quickly that someone is trying to compromise our birds, we've paid for the sensor suite. And it gives us time to turn things off, reposition the spacecraft, or do other things—actions that we, the operators, can take on our own. But then we need to share that information. Maybe set up a joint military-civil-commercial alert center . . ."

"We already have one—the Space Control Center," Aster interjected.

"Yeah, but we need to set up ways for commercial people to automatically send alerts to the SCC," Molinero pressed. Aster nodded, arms folded now as he paced the stage.

"And the question, General: An *Aviation Week* article a few years ago said we already have a super-secret, two-stage-to-orbit spaceplane of some sort that could be used for space control. Isn't it in service today?" Molinero asked, then sat down. The auditorium hushed again.

Aster stopped, arms still folded as he stared at the floor. He said nothing for a full twenty seconds, then gave Molinero a sharp glance and clipped, "I can't get into that here."

Hands on hips again, he signaled that the hotwash session had ended. "Ladies and gentlemen, this concludes the DEADSATS II wargame. Thanks to all of you for your time and very valuable input. Now let's get back to work. We still have serious situations on our hands."

Yeah. Like restoring lost space resources before the whole damned world explodes, Aster thought as he left the stage. *Speed better be making headway with that spooky Blackstar system!*

16 APRIL/DAWN: ISRAELI DEFENSE FORCES OUTPOST, NEAR KIRYAT SHMONA, ISRAEL

The SPECTRE-1 unit was hardly compact. In fact, it was a hell of a lot bigger than Brigadier General Mike Fisher would have liked, which made it less mobile and more difficult to conceal. The Army one-star had badgered its Raytheon developers for years, pressuring them to build a truly field-deployable version. But the laws of physics and engineering, aggravated by the constraints of a brutal development schedule, conspired against him. Finally, the need to get it into a twenty-first-century combat environment forced him to accept this huge prototype.

He finally conceded that getting SPECTRE-1 into the field and *proving* it could counter bad-guy weapon systems was absolutely essential, or Pentagon beancounters would kill the program. Deploying an oversized prototype was preferable to endless tinkering and refinement in the lab, hoping the next iteration would be the ideal. *Sometimes, you have to shoot the engineer and fly the damned airplane!* he'd fumed to the Raytheon program manager, who begged for more time to improve its design. So, he was finally here, risking the loss of SPECTRE-1, his own brainchild—an information-warfare dream come true. But the potential payoff of this particular mission far outweighed the considerable risks, an argument he'd finally won in Washington.

Dark skies had threatened rain all night, but the showers never came. Instead, the thickening gray, scudding clouds miraculously gave way to clear skies at the first light of day, giving Fisher's small unit of Golani Brigade soldiers, several Shin Bet observers, and a no-name American "spook" a magnificent sunrise. That same dawn was shared with Lebanese villagers a few hundred feet beyond the border, of course. The Army officer carefully surveyed their collection of shabby buildings and the desert landscape around them, using powerful light-gathering binoculars.

Yes, being this close to the border certainly is dangerous, Fisher admitted to himself silently. He could sense the presence of Hezbollah fighters in the neighborhood, hardened eyes watching Fisher's team from dark windows and, probably, from underground bunkers. Those renegades would be searching for any opportunity to dart across the border, covered by a protective wall of automatic rifle fire and rocket-propelled grenades. Such attacks on Israeli outposts were far too common, he knew. Even one or two Golani casualties would be headline news on al Jazeera and CNN, especially if one of the ever-present radical Islamist cameramen could capture a few images of Israelis being slain. Those images definitely would appear on an

Internet website, regardless of whether al Jazeera or CNN aired them.

Still, bringing SPECTRE-1 here, to the Israel-Lebanon border, was worth the risk, Fisher thought. Although this wasn't officially his war, the Hezbollah incursions of 2006 had greatly increased Israeli interest in the novel system. SPECTRE-1 *was* his concept, however, and the Israelis now were keenly supportive of its development. Despite the Pentagon's official displeasure at Fisher's decision to come here, so he could personally obtain irrefutable proof that SPECTRE worked in a real-world combat environment, he'd managed to convince Paul Vandergrift that a joint American-Israeli mission was essential.

Because these were troubled times for Western civilization, the National Security chief was easily persuaded that his Army liaison officer—and Fisher's sophisticated information ops device—could be a critical "silver bullet" weapon in the years-long, black-ops battle with Iran. Neither Vandergrift nor Fisher was about to let raised Pentagon eyebrows prevent them from learning how SPECTRE performed in the field, and from garnering yet another kudo from President Boyer, as a bonus. Fisher knew that his boss, Vandergrift, desperately needed to elevate *his* own capital with the Oval Office, and this mission could do exactly that. *And maybe, just maybe, also get a second star for Mr. Vandergrift's Army liaison, too.* Fisher smiled, still scanning the softly lit scene before him.

His team had been at the border for three nights. Hundreds of eyes that usually watched flocks of sheep in southern Lebanon had obviously spotted it. CIA messages forwarded to him said the Syrians also had been alerted to the presence of his odd-looking, heavy-duty Toyota pickup with the custom camper top. The truck had been carefully tracked as it drove along the rough non-road tracks that separated Israel from Lebanon. Even now, Fisher knew his team was being observed.

Would Hezbollah ever oblige him by sending an un-manned aerial vehicle overhead to get a better look at the Toyota and his team? *Just one UAV, a pilotless drone. That's all I need. Come on, suckers. Put one in the air!* Just one UAV would allow him to prove SCEPTRE's miniradar could detect a diminutive recon drone, a tiny, stealthy, remotely flown aircraft made of carbon-fiber composite material. Then he'd unleash SPECTRE's big gun. For five years, he'd personally campaigned for the secret development of information operations or "IO" technology capable of penetrating the guidance system of airborne enemy drones, then taking control of the birds.

Once inside the drone's electronic brain, SCEPTRE-1 could siphon off any video images the UAV might be beaming to a ground station, revealing the location and identity of its controllers in the process. Just ID'ing the drone's launch and control station would be a perfectly satisfactory trophy for this initial go-round. And the best place to test SPECTRE in a moderate-threat situation was on the Israel-Lebanon border, where Iranian-built UAVs were routinely deployed by Hezbollah.

The damned mini-UAVs were causing more than minor concern in Tel Aviv, too. Sure, they might look harmless, but let them fall into Hamas' hands, with Quassam- and Katyusha-rocket teams positioned nearby, and the entire Israeli population in this region would would be at risk. No wonder the IDF had jumped at a chance to help Fisher test his SPECTRE UAV-control unit, especially after he proposed a joint development program.

The Golani squad members guarding Fisher and the Army specialist-operator, who had accompanied the American general, were not particularly friendly. They were battle-hardened professionals, acutely aware that they were perched too close to the border. With no protective armor, they were sitting ducks, easy targets for a rapid-fire barrage of rocket propelled grenades from those scruffy houses across the border, an imaginary line in the sand where UNFIL feared to tread. Even one Golani casualty could be a hardship for Israel, a tiny country with a very small pool of eligible draftees. Yes, every IDF soldier was accustomed to being outnumbered three or four to one and still coming through a pitched firefight without a scratch. Israelis fought smart and fought carefully, but Hezbollah also had battle-hardened fighters who readily soaked up casualties and kept on fighting, as the world saw in 2006.

In the early-dawn gloom, the Golanis did their jobs in complete silence. So near the border, they didn't dare make noise. Farmers on the other side had ears better than any fancy unattended acoustic sensor. A dropped canteen cup or hushed expletive or the snap of a clip being seated in its weapon damn sure didn't sound like a sheep or a goat. Almost any unusual noise would prompt those farmers to dispatch their children to a neighboring village, warning that the Israelis were coming. Then armed Hezbollah units would materialize, leaving underground tunnels to swiftly strike the small unit. That was the norm here, an outpost outside Kiryat Shmona, a town of bomb shelters, destroyed houses, and bullet-pocked walls.

It suddenly seemed very far from Mike Fisher's hometown of Elkins Park, just outside of Philly, one of many burgs along the SEPTA Main Line. A Penn State electronics engineer, Fisher had begged his way into Army ROTC after a borderline performance on the physical exams. Officers who ran the ROTC unit had liked Fisher and his bulldog approach to problem solving, though; he had an ability to hit a mathematical or technological wall, yet keep battering away until he found a solution. After graduation, he'd turned down a quick-turn graduate degree in engineering—which is what the Army clearly wanted—and hounded the service to send him to Wharton, instead. There, he honed those natural bulldog talents into a no-nonsense, business-smart dynamo.

Once he was on active duty, Fisher had quickly become a bureaucratic snowplow, an officer with a reputation for getting things done, paperwork be damned. No ROTC instructor had believed the hard-headed cadet would ever wear stars or force his way into the Pentagon's heady halls of power, but he had. With a Wharton MBA in his back pocket, Fisher blazed through the ranks, pushing aside every obstacle the old-boy Army network plunked in his path, until it threw up its hands and gave him a star, a small desk in the Pentagon's inner-ring

"Siberia," and orders to stay out of everyone's hair. Some full general had hoped the promotion would prompt Fisher to shut up about "transforming the Army" and retire early, then take a cushy senior management job with a defense contractor. But that was not Mike Fisher's plan.

From the moment he saw the old *Star Trek: The Wrath of Khan* movie, he'd been obsessed with the idea of a device that could surreptitiously take control of the enemy's weapon systems. "Just imagine," he'd enthuse to any sympathetic colleague he could buttonhole, "a system that could get inside the enemy's reconnaissance platforms, capture the bad guy's imagery and his signals intelligence, and even retarget his missiles. Hell, we could fly his missiles and UAVs right back to their source!"

Assigned to Fort Dix, then-Lieutenant Fisher built a rudimentary version of what eventually would become the SPECTRE info-ops "computer attack" system. Logically, from his point of view, he decided to test it on an unsuspecting New Jersey state trooper manning a radar trap along Mount Holly Road. Unsuspecting speeding soldiers were an easy way to beef up a trooper's ticket quota, which made the police radar system a viable "enemy" target, the young officer concluded.

Fisher's crude IO-attack system captured the trooper's incoming radar signals, amplified the pulses to an intense level and fired them back to the police car. The trooper's receiver basically blew up, driving its auto-calibration feature offscale, rendering the system legally blind. Later, Fisher modified his device to return a false series of radar pulses, which made the speeding car appear, on the radar, to be crawling instead of flashing by. Some admirable tech-police sleuthing busted Fisher and got him in deep trouble with a no-sense-of-humor Fort Dix commander, but the caper proved the SCEPTRE concept was technically feasible. That, in turn, led to some serious Army funding and twenty years of development.

Now he was in Israel, still relatively young, still a great senior management candidate for one of those defense contractors in El Segundo, but in the company of a local IDF detachment. And those Israelis were *not* happy about babysitting an overgrown kid with a super gizmo, even if he was a Yank general. At the end of the day, sure, Fisher was a soldier, and one of them. But the guy was still weird, they'd concluded. The night before, in whispers, they'd asked him, "You ever eat a cheesesteak?"

"No," he confided. "My parents kept kosher." He paused, maybe wondering whether his maker would greet him on the morrow. "Well . . . once I did . . . okay, twice. But not on high holidays!" And the soldiers had laughed. Some promised that they, too, would go to America and taste the forbidden food.

Suddenly, Fisher's screen flashed, echoed by a verbal alert from the Army specialist-operator at the general's back, who monitored a duplicate screen. An image, a small blip, had appeared, heading toward their position. *The bastards launched a drone!* That meant Hezbollah was nearby. The general flashed a thumbs-up to the short-haired Shin Bet guys outside the camper, who relayed the alert via hand signals to the IDF soldiers. The latter quickly dropped into their sandbagged defensive positions, sighting along snub-nosed weapons, waiting for a still-invisible enemy.

"Wait for it," Mike cautioned under his breath, moving a captain's bar-shaped cursor over the blip, then tapping a key to switch the small radar from surveillance to lock-on and auto-track mode. He would not activate the super-secret control beam until he could see "the whites of its CCD-imaging eyes," as he had promised Vandergrift in the White House and God-knows-how-many generals at the Pentagon.

Hunkered inside the camper, Fisher was gripped with excitement. He began counting, then, under his breath, declared: *"Now!"* A key-tap activated SPECTRE's targeting subsystem, firing a beam of covert, low-probability-of-intercept ultrawideband pulses. A green icon appeared in the corner of his laptop's screen.

"Hot damn! We're in, sir!" the young specialist whispered loudly. Fisher ignored him and focused on the display, slowly moving a tiny joystick to the left. The now-winged blip moved to its left, in response. A gentle bank-turn brought the blip—

Hezbollah's unsuspecting surveillance drone—slowly around 180 degrees, half a circle. Unasked, the Army specialist-operator pulled up a grid map of the Israel-Lebanon border area, then zoomed in, expanding the zone as if it were an Internet video game. The slaved computers ensured that the same map appeared on Fisher's screen, improving his sense of the UAV's location in relation to the Toyota and SPECTRE.

Okay, don't telegraph what we're doing, Fisher talked to himself, silently. *Don't show 'em what we've got. Let SPECTRE work its magic.* A few key strokes activated the attack system's preprogrammed protocol, a sophisticated set of maneuvers that would soon lure the enemy UAV into Israeli territory—and into U.S. Army possession. Fisher held both hands inches above the keyboard, eyes locked on the screen, holding his breath. The blip that represented the UAV suddenly banked sharply left again, then dived at the ground as it swung back toward the border, accelerating in the process. Scant feet above the ground, the tiny aircraft leveled off, its engine now audibly whining, straining for speed, as the craft zoomed across the invisible border. Fisher jumped up and out the camper's door in time to see the drone pass near the Toyota.

"Damn it, sir! We did it! We stole the friggin' ragheads' UAV!" The specialist was at Fisher's shoulder, beside himself with glee, the odor of stale coffee preceding him.

Fisher barked a quick order, sending the specialist and a Shin Bet technician racing after the UAV as it settled to the ground, skidding on its belly. The small airplane caught a wingtip in the windblown, rocky soil, spinning the small craft in a cloud of dust. It stopped, still intact, just before the two men skidded up to it. They each grabbed a wing and turned toward Fisher, who had stepped away from the camper, keeping the hulking vehicle between him and the border.

The camper exploded, pitching the truck into the air and its cargo to the side, narrowly missing Fisher. "RPG!" he shouted, instinctively flattening himself in the dirt as shrapnel and debris from the camper whistled overhead.

Between bursts of automatic-weapons fire, the general

heard someone screaming orders, in Hebrew, into a field radio. The distinctive clack of a Kalashnikov reached him, punctuated by dozens of explosions. Other weapons opened up a few feet from where he lay, their deafening roar assuring him the Israeli soldiers were still alive—and had plenty of targets. Fisher stumbled to his feet, clawed for a hip-mounted 9-mm Glock, but came up empty-handed. His escorts had insisted that only they carry weapons; they refused to have an armed American general, a man who had never been under fire, at their backs. Ears ringing from the first RPG blast, Fisher knew he was in shock, but the bulldog deep within arose, forcing him into a lead-footed slog, headed for the downed UAV. Flying grit stung his skin and eyes, but, through the swirling dust, he glimpsed a desert-camouflaged lump near one of the drone's wingtips. It was the GI, his SPECTRE operator.

He's down! Fisher had to help his soldier! But he staggered like a stupefied drunk, stumbling through a nightmare at snail-speed, trying to reach the downed GI. Distantly, he was vaguely aware of the Shin Bet technician kneeling near the UAV's other wingtip, firing rapidly with a two-handed grip on an automatic weapon. The man's mouth was open, apparently screaming something, but Fisher only heard the cacophony of gunfire and explosions.

Somthing slammed him in the back, sprawling Fisher forward, facedown in the dirt. Before he could get a hand down, charcoal-faced figures with heavy beards and moustaches surrounded him. *Hezbollah!*

"No! Not SPECTRE!" Mike screamed. Illogical, perhaps, but that was his first thought. He twisted free of one fighter's grip, rolled onto his back and kicked a second bearded man in the face. "I will fuckin' die before you get your hands on it!" he yelled, scrambling to his feet, then diving on the man he'd kicked. Fisher may have passed for a stereotypical geek, but he was a trained soldier, now engaged in hand-to-hand combat. He wrestled the rifle from his victim, slammed its butt into the guy's beard, then spun toward the other would-be captor and started firing.

That Hezbollah fighter exploded in blood as Fisher's first burst ripped through his face. Then a second went down to another spray of slugs. But before he could spin again, Fisher was slammed to the ground, pinned by a knee jammed in his back. Fisher kicked backward, but an unseen guerrilla, skilled in winner-take-all street combat, swung a boot into the general's testicles. Excruciating pain turned Fisher's world white, and he instantly puked, just as his face was jammed into the dust by the hulk on his back. The American was losing consciousness, but he recognized Arabic being shouted. A sharp pain under his shoulder blade was followed by the odor of blood. He gagged, coughed and saw blood spew from his own mouth. Fading out, he had an odd sense of satisfaction that he was still wearing the black beret of the U.S. Army. He whispered a prayer as the world grayed, then darkened, only vaguely aware of being dragged away.

A string of 20-mm shells ripped into the earth along the border, but two hunched-over Hezbollah fighters never broke stride, dragging the American between them. Two others shadowed them, Kalashnikovs raking the air on either side of the small team. Then the entire landscape shook with the impact of a 500-pound bomb. A pair of F-16 fighters screamed above the invisible border, their 20-mm cannon fire again stitching the ground, spewing a wall of flying dirt and rock. Artillery pounded in the distance, accompanied by the deadly scream of shells overhead. They exploded on the Lebanon side.

The Golani soldiers never stopped firing. Anything that moved in front of them went down. A well-placed round dropped one of the fighters dragging Mike Fisher, but an answering fusillade of lead forced the Israelis to duck behind their sandbags momentarily. In seconds, the Hezbollah team had dragged Fisher across the border and into a slight depression, shielding them from Golani fire.

"We can still get him," a grim-faced IDF soldier yelled to a Shin Bet handler. The latter turned to the only remaining American, who was *not* wearing a uniform or any identifying insignia.

"No, we can't," the American said quickly. "Get your unit

out of here." Protected by covering fire, the team retreated in leapfrog fashion—one group running, then dropping to a knee—while the other cluster fired at the Hezbollah terrorists. A second pass by the F-16s triggered two massive explosions. The quarter-ton Mark-82 bombs gave the Israelis a few needed moments to hook a Humvee on the twisted, shattered, partly melted Toyota and drag it away from the border.

Mike Fisher's family would be informed that he "was killed in action, on a classified mission of critical national security. No, we can't say where, or what he was doing." That would be the official story. That an American general had been captured by Hezbollah fighters would never be known. Until someone whispered the news into Dagger's ear.

10

SPACEPLANE

Barely into "near-space," that thinning, in-between region of Earth's atmosphere, the SR-3 "mothership" cruised comfortably. Speed Griffin cross-checked his XOV-2 space-plane's data readouts against the SR-3 copilot's figures coming over the fiber-optic link connecting the two vehicles. Because his spaceplane was nestled under the mothership's belly, Speed could see nothing outside. The vehicle's thick windscreen remained shielded by a shaped, conformal fairing that blended the SR-3's fuselage with that of the XOV, pre-venting supersonic air from prying the two vehicles apart. Without that protective fairing, repeated standing shock waves—each changing shape as Mach numbers increased—would cause undue stress on the spaceplane and its mother-ship.

"Initiating accel in five . . . four . . . three . . . two . . . one. Mach 1.5," the copilot's monotone voice announced through Speed's earpiece.

Griffin verified he'd done everything that could be done in preparation for his imminent separation from the SR-3. Listening to the copilot's speed readouts, Griffin mentally scanned the spaceplane from front to back, and from top to bottom, a practice he'd used throughout his illustrious flying career. *Never know what you'll turn up,* he thought. Nose to cockpit: all okay. Underbelly: gear and skid up. Top of fuselage: canopy, latched and secure. Q-bay: secure.

The small Q-bay, or payload cavity, behind his cockpit carried a package of small, second-generation "Tacsats"—tactical satellites—nested together. Not much he could do about them. Either they were mounted securely on the trapeze system that would extend and disperse them in orbit, or they weren't. Q-bay doors: closed and locked, as verified by the green PAYLOAD light on his right console. Tail section: obviously secure, or he and the spaceplane wouldn't be here, suspended beneath the SR-3's belly. The XOV's short-and-thick vertical tail, which ran for some distance along the XOV's spine, provided a modicum of directional stability when flying in the atmosphere. But its actual purpose was to serve as a pylon that attached his spaceplane to the mothership.

"Mach 2.0; all Mama systems 'go,'" the SR-3 copilot droned. "Launch in nine-zero seconds."

Griffin thumbed a mike button on the throttle in his left hand. "Gaspipe's a 'go.' Mission control, comm check. How're we lookin'?" Miles away, a team of engineers huddled in a small, windowless room scanned banks of computer displays blanketed with graphics, alphanumeric data, and colored bar-graphs at a super-secret "nonexistent" location in northern Nevada.

"You're five-by. Mission Control's a 'go'; everything's in the green. Good luck, Speed," radioed a Lockheed Martin flight test engineer at the Air Force's secret Groom Lake facility, known to generations of test pilots as simply "The Site." All transmissions were encrypted to make sure an eavesdropper would hear only garbled "fuzz" on the dedicated Blackstar mission frequency.

Long-dormant butterflies took flight in Speed's gastrointestinal tract, just as they always did prior to an XOV launch. A few more were probably airborne today, though. This was the XOV-2's first flight in more than a decade, and Murphy had a way of coming up with new tricks, despite meticulous preflight preparations by dozens of experts. Speed had spent hours in the XOV-2 simulator at Gator airstrip in Florida, sweating through every potential emergency conceivable, re-familiarizing himself with the quirks of flying a one-of-a-kind vehicle outside the Earth's atmosphere, then back to a runway landing. Still, you wondered. Will the new batch of gel-type fuel light off okay? It passed vacuum chamber tests, but will it pump and flow properly, with no bubbles or cavities that might cause a bank of thrusters to blow out?

At some point, you just had to trust the damned system, assuming it would work. God, how he hated the waiting! That's when the "what-if" doubts crept in, threatening to steal a pilot's nerve, upstaging the technical know-how and cool demeanor that ensured you'd perform flawlessly. Once the drop happened, you're always too busy to sweat anything but doing the job. Fear of death? Not really. The real fear is always the same: a fear of screwing up and letting your team down. *Let's fly this SOB!*

"Mach 2.7; Mama's still a 'go.' Standby for launch . . ."

"Gaspipe's a 'go.' Purging engines . . . now! Fuel tanks pressurized." Griffin pressed a switch, knowing super-cold nitrogen gas from an onboard storage bottle was blasting through the plumbing and combustors that constituted four banks of aerospike engines embedded in the blunt, rectangular aft face of his soon-to-be spaceship. Another bottle dumped its pressure into the labyrinth of fuel tanks, pressing toothpaste-like gel into stainless-steel lines and against a handful of still-closed valves. The engine-health display on his right flat-panel screen showed every parameter in the green. Ready to fire.

Only seconds to go. Time for those last items on his checklist. Suit pressure: *Auto.* Confirmed, he noted silently. His pressure suit would inflate only if the XOV's cockpit started leaking air. Visor: *Down.* He reached up and lowered a gold-tinted visor on the astronaut-type helmet that was now twist-sealed into the circular neck-fitting of his orange spacesuit. It was the same type of pressure suit that had been worn by myriad SR-71 Blackbird pilots and space shuttle astronauts. Griffin squirmed slightly—his butt had to be securely strapped into the conformal seat—then took a firm grip on the XOV's controls, throttle in left hand, sidestick controller in right. He took a deep breath, then exhaled slowly. *Show time!*

"Mach 3.0. Launch in five, four, three, two, one. *Release!*"

Griffin felt the thump of jawlike clevices opening, then his confined world dropped like the proverbial rock. Instant weightlessness and a blast of brilliant sunlight greeted him. Seat straps held his body and spaceplane together, and the gold-colored visor shielded his eyes. *One-thousand-one,* he whispered. That one-count gave the XOV time to fall a comfortable distance below the SR-3 mothership. He jammed a press-and-hold button on the inside of the throttle. It was answered by an instant kick as the aerospike engines ignited, slamming his head against the seat's headrest. *Shit! Forgot about that!* The thought was clarity itself, but only a tiny flash across his consciousness.

Eyes locked on the engine screen, Griffin was greeted with a comforting rise of vertical "tapes," indicating that all four banks of aerospikes were firing equally, fed by that boron-based gel fuel and atmospheric oxygen. He glanced outside and nudged the control stick forward to arrest a slight climb. He knew the SR-3's pilot would have banked to the left immediately after the drop, but, just in case, Speed didn't want to risk colliding with the huge mothership from below.

"Mama's clear. Tally on Gaspipe," radioed the SR-3's copilot, indicating the mothership had turned away, and its crew now had the XOV in sight.

"Gaspipe's clear; accelerating to pick up climb schedule," Griffin responded, easing the throttle forward. Airspeed increasing rapidly. Mach 3.5; 59,700+ feet; all engines firing, producing thousands of pounds of thrust and pressing his back firmly into the seat. Bit more nose-down stick pressure, then bump, bump with the right thumb to trim out control stick forces that kept trying to pitch the nose up as speed increased. Sight picture out the front's perfect for level acceleration. Heading looks good.

"Rog, Gaspipe. Mission control shows you all-green. Cleared for climb."

Speed smoothly pulled on the stick in his right hand and pushed the throttle to its forward stop with the other. He noted the Earth's distinct curvature as the horizon disappeared under his XOV's nose. A deep-purple sky quickly darkened as the spaceplane rocketed toward space. Griffin watched the attitude indicator symbol on his primary flight display "roll" toward him until the 80-degree nose-up pitch bar approached the center horizontal line. He was now on his back, looking almost straight up into the black abyss. Slight forward pressure arrested the pitch maneuver, steadying the XOV-2's climb angle. Mach numbers were clicking by faster than he could read: six, seven, eight. The powerful engines' thrust created accelerations of several times his own weight—so-called g-forces—steadily pressing on his chest and making it hard to breathe.

"Gaspipe, you're on the zoom profile. Lookin' good, Speed," the flight test engineer said. Griffin was conscious of a throaty rumble behind him and the comforting vibration of powerful aerospike rocket engines pushing his vehicle higher, faster. His eyes remained inside the cockpit now, scanning the electronic displays. Engines good; all parameters in the green. A Mach number-versus-altitude graphic placed the small symbol that represented his XOV smack on a curved magenta line in the corner of that primary screen. *Hold what ya got. Steady . . . You're not pitching over onto your back. Just the inner ear's response to rapid acceleration. . . .*

Griffin stole a quick left-right glance outside. Brilliant sunlight filled one side of his wraparound windscreen; the other was stone-cold black. Eyes back inside, he saw the altitude tape zip past the 100,000-foot mark, still climbing. A few more seconds and the Mach indicator would switch to feet-per-second readings as the outside air thinned. At some point, there weren't enough molecules of air for "Mach number" to be meaningful as a relative-speed indicator. "Airspeed" and "Mach" became irrelevant terms once you reached the vacuum of space. Still on the magenta flight-profile line, still climbing.

"Gaspipe, Control. Give me two left, Speed. You're a hair off inclination."

"Rog. Two left." The SR-3 had launched him on a heading that should put his spaceplane into the correct orbital inclination, about a 35 degree angle with the Earth's equator. But it was his job to place the XOV precisely into that orbital inclination. If he didn't, those nanosats in the Q-bay wouldn't be inserted into the correct orbit, which would preclude their getting the right intel imagery, and . . . yada, yada . . . *Just fly the purple road to space . . .*

"Gaspipe, Control: twenty seconds to engine cutoff . . . *Mark!*"

Griffin acknowledged, then concentrated on holding the vehicle precisely on its climb profile. Too low and he'd never make the intended orbital altitude after the engines stopped firing. He'd have to dump the nanosats among those scattered air molecules of Earth's very high upper atmosphere. What little air was still there eventually would create enough drag to slow the nanosats' speed, letting them fall back into the atmosphere and burn up long before their programmed life expectancy. Not good. Too high and they'd be in the wrong elliptical orbit to take pictures of what the intel folks and military commanders down there needed to see. Equally not good. *Don't screw this up. Right . . . there . . . engage autonav.* Once he'd tweaked the flight profile to a human's best abilities, it was time to turn control over to "George," the spaceplane's autopilot.

Years before, the XOV team had worked through the classic robot-versus-human tradeoffs. The flight control team argued that a computer-controlled autopilot/navigation or "autonav" system could fly a tighter, more precise three-dimensional profile into space than a human pilot could. After all, computers could sample flight-profile data thousands of times a second, then adjust control surfaces and thrusters precisely, nudging the XOV into exactly the right place. Their logic had won the debate until the first test flight into orbit. Speed had damned near lost consciousness when the XOV, flying hands-off and controlled by the autonav system, started porpoising in pitch and ratcheting back and forth in roll with increasing excursions and rates. He'd worked frantically to disengage the system, then flew the vehicle manually until control was regained. A lot of engineers on the ground almost needed diaper changes that day. As did the spaceplane pilot.

Post-flight analyses had showed an unexpected latency between computer commands and the flight control system's actual response. That small time delay caused the computer to decide nothing was happening, so it ordered *more* of an input, just before the first command had been carried out by the flight controls. In short, computer commands and control system responses were out of phase just enough to turn the XOV's flight profile into an E-ticket-ride, a cross between a rodeo bronc and a roller-coaster. Of course, subsequent work by smart, diligent engineers ultimately resolved the flight-control latency problem. But since that first flight, the agreed procedure was to have a pilot always keep one finger on the "Alpha-Sierra"—or "Aw-Shit"—button, the control stick's autonav-disconnect switch, whenever the automatic system was engaged. Today, no sweat. The autonav assumed control with flawless precision. No Alpha-Sierra rides today.

Griffin continued his careful instrument scan until the engines automatically ceased firing. The sudden lack of acceleration threw him forward, retention straps biting into his shoulders and waist momentarily. Then silence and an otherworldly sense of peaceful floating commanded the pilot's consciousness, a marked change from the g-forces that had been trying to crush his body only moments earlier.

"Gaspipe, Control shows successful engine cutoff on schedule."

"Confirmed, Control. Attitude cross-check?"

"You're right on target, Speed. Couldn't be in better position. Nice work!" the engineer radioed. Griffin could hear faint cheering in the background.

"Thank *you*, Control. Log another successful XOV launch. Nice to be back up here."

The pilot stole a moment to revel in the weightlessness of space flight. Sure, he was still tied into the seat, but his legs and arms wanted to float, and he could feel blood gravitating to his upper body, making his face bloat a bit. He tilted his head back, noting that he was now flying upside down in relation to the Earth. Strange to look *up* in order to see the ground *down* there, he thought absently.

Okay. Get to work. At this relatively low altitude, barely into space, a complete orbit around the world didn't take long. Griffin ran through a payload prelaunch check, confirming the nanosats' onboard power systems were activated, but in standby, until deployment. Mission Control, monitoring the payload's telemetry signals, verified all parameters were "nominal," then cleared the pilot for Q-bay opening at his discretion. Speed scanned the cockpit, made sure everything was in order, then commanded the bay's doors open.

"Control, I show Q-bay doors full-open. Confirm."

"Rog, Gaspipe. Doors full-open. Cleared for release."

Griffin took another deep breath and activated the "Payload Extend" command, unfolding a trapeze-like structure from the Q-bay. In a narrow, automobile-type rearview mirror mounted on the canopy's upper frame, the pilot could see a sliver of that action behind him. The nanosat package was still firmly mounted on the now-extended support. Collectively about the size of a large garbage can, a dozen small satellites were nested together like a stack of shallow Dixie cups.

Speed selected the DEPLOY icon on a touch-sensitive display and watched the nest slowly drift away from the XOV, propelled by a simple mechanical spring in the mounting rack. He bumped his controls and felt tiny thrusters embedded along the periphery of the XOV's tail puff momentarily. Just enough to add energy and accelerate the spaceplane a few feet per second, which served to raise his orbital altitude some sixty feet. In the mind-bending world of orbital mechanics, going faster translated to moving higher. It now appeared that the nanosat package was drifting away from him.

He carefully maneuvered until the spaceplane was station-keeping in position above and behind the nanosat package. Still flying upside down with reference to the Earth, Speed could see the can-like nested structure between him and the planet. *That's more like it. Now I can see what the hell's going on out there. Wow! What a sight!* he marveled silently, reflecting that very few people had seen their world from this unique perspective.

"Control, Gaspipe. Package released, and I'm a good 100 meters above and behind it now. Securing Q-bay."

"Rog. We show a clean release and the package in powered standby. We'll wait for your clearance to activate, Speed."

Moving deliberately, Griffin commanded the trapeze-like structure to fold and stow itself in the Q-bay, then the cavity's doors to close. This was one of the most critical procedures of the entire mission. If that Q-bay door wouldn't close properly, a gap would be left in the XOV's smooth exterior. Some computer models said that a *small* gap in the bay doors could be tolerated during reentry to Earth's atmosphere. But a few minor deviations in reentry velocity, or a slight roll at the wrong time, and the gap could become an entry point for white-hot plasma to create a blowtorch effect and burn through the XOV's structure. Seven years earlier, America had lost seven astronauts when something akin to that blowtorch effect had melted space shuttle Columbia's skin and vulnerable substructure.

Bottom line: *Make sure the damned Q-bay doors are closed, Griffin*. Of course, he couldn't see whether the doors were completely closed, even in his mirror. Still, he saw no exposed edges, and the green "Q-Bay Closed" light was on, so he assumed the doors were closed. One step closer to home.

"Control, Gaspipe's buttoned up. Clear to deploy sat-package," Griffin radioed.

"Rog. Control's activating deployment sequence." Griffin watched as the nested formation of small satellites separated, then unfolded as individual spacecraft. He was amazed at how much stuff clever engineers could fold into such a small volume. He marveled at the origami-like structures bending, twisting, always growing until, finally, they snapped into stable vehicles, each armed with two wing-like solar panels. The average person, viewing those six vehicles now hovering in space, would never believe they'd been carried aloft in the small, now empty compartment a few feet behind Speed's cockpit.

"Control, Gaspipe has six—repeat, six—subpackages dispersed. All look good from here."

"Thanks, Gaspipe. Everything's nominal here, as well. Clear to separate."

Griffin's primary task was complete. He'd delivered the nanosat constellation to orbit. The ground crew wizards would take it from here, activating a sequence of commands that would slowly reposition the six tiny platforms until they were a certain distance apart. Their control systems, relying on accurate laser-rangers that precisely measured their separation, would station-keep the six tiny satellites in a precise 3D geometry. Through the magic of sophisticated software, this "distributed" system of six platforms—each configured with a high-resolution imaging system operating in both visible and infrared wavelengths—would act as a large-aperture spy satellite. Their individual images would overlap slightly, enabling processing algorithms to literally stitch together a seamless, high-resolution photo of a ground target. Not as high-res as that K-15 imaging-sat those damned druggies had killed with their maser, but good enough for critical near-term intelligence needs.

Griffin completed his post-deployment checks and started setting up for reentry. He'd now have to rely on computers to calculate precisely when and where to somersault the XOV, point its tail down-track, then fire the retrorockets to slow his spaceplane enough to reenter the atmosphere at the proper angle and speed. He could manually override the system, if need be, but the odds were that a human being would foul something up. Reentries were tricky. Too fast or shallow and you could skip off the atmosphere and back into space. Too slow and the reentry angle would be so steep your spacecraft would burn. And hitting the right time-distance point was especially critical if you needed to land at a specific location.

A few minutes later, the XOV-2 fired its retros, slowed sufficiently, flipped back to a belly-forward and -down attitude, and started descending into the atmosphere. Griffin really didn't appreciate this part of space flight, because there wasn't much a pilot could do but hang on and ride it out. He was no longer in control. Soon, faint tongues of orange-red plasma licked back from the XOV's highly swept edges, creating an eerie, Halloweenish glow inside the cockpit, growing brighter with each passing second. That now-ionized layer of super-heated air surrounding the vehicle also prevented him from communicating with the mission controller. Until he slowed sufficiently, no radio-frequency signals could penetrate that ion shield.

Griffin didn't feel the heat, but he knew it was there, threatening to burn his small world to a metal-carbon crisp. He knew exactly what to expect, having been through this phase many times before. But . . . this critter was ten years older now. Did anybody *really* know the long-term aging characteristics of exotic alloys that made up the structure of his spaceplane, especially after being heated and stressed repeatedly by the buffeting and vibration of numerous reentries, then sitting in the barn for years?

Speaking of which, airframe buffeting had begun, building slowly, steadily. More fire outside, more vibration and buffeting inside. He could still read the rapidly shaking instruments, but the fillings in his teeth felt like they could jar loose at any moment. G-forces were now increasing, the thickening outside air's resistance trying to force him through the seat again. Not too demanding, but the sustained 3-gs or so made you work, tightening legs and abdomen to keep the blood up in your torso and cranium. If his brain were starved of blood-transported, life-giving oxygen, a pilot would first gray out, then black out completely. Passing out wasn't a good idea when flying the nation's one and only manned spaceplane.

Then reentry was over. The flames disappeared, vibration ceased, and the XOV-2 reverted to being an airplane again. *More like a low lift-over-drag glider—or a brick*, Griffin reflected, scanning the instruments and watching the nose pitch down.

"Control, Gaspipe's back with you. Airspeed's alive. Resuming manual control."

"Roger, Gaspipe," the controller said, his tone reflecting a roomful of relief that communications had been reestablished. The nail-biting reentry was behind them. "You gonna fly this puppy all the way down, Speed?"

"Look, I only get a few minutes of actual stick time per sortie. You bet I'm flying it home!"

"Roger that. Corridor's cleared; commence parameter-ID maneuvers at your discretion."

Griffin had campaigned for including a few flight test-type maneuvers to add more aerodynamic data to an ever-growing knowledge base. Those data could, eventually, improve the XOV's mission effectiveness. There was less than an hour's worth of post-reentry aero data on the books, so far, and every new tidbit taught the engineers and pilots a little more. He quickly put the spaceplane-turned-airplane through a series of pitch, roll, yaw, and other maneuvers designed to acquire both performance and stability-and-control data. Then it was time to prepare for landing.

He would touch down on a long runway at the infamous Area 51, or Groom Lake, also called "The Ranch," in northern Nevada. No matter what the latest crop of curious black-aircraft chasers called it, the facility rigorously guarded its secrets and ignored the rumors, even decades after the first of Kelly Johnson's highly classified Lockheed "Skunk Works" aircraft landed on the hard-surface dry lakebed there.

"Gaspipe's through 70, descending. APU on," Griffin clipped.

"Looking good, Gaspipe. On profile; all green. Weather's clear, surface winds light and variable. Cleared to land."

Still hundreds of miles from the secret runway, yet he was "cleared to land." That's spaceplane flying for you; normal distances were almost meaningless. Griffin could see much of southern California below as he banked, then Las Vegas rushed at him. Soon, he was over the expansive Nellis AFB test and training ranges, still at high altitude, but descending rapidly toward the dry desert terrain.

He entered the circling pattern now depicted on his primary guidance display, cross-checking speed and altitude.

With no engine thrust, there would be no go-around if he screwed up the approach. Either he put the XOV right on the runway, at the right speed, or he landed in the desert—at one end of the runway or the other. *Okay, do this right, Speed.* He always talked to himself through landing, a quirk that had unnerved more than one backseater in F-16s and F-15s, years ago. But the self-dialogue focused his attention.

Steady, steady . . . Gear down. Three green. Flare slightly, a bit more. Hold it, hold it . . . Griffin felt the two wingtip- and single centerline fuselage-mounted skids squeak, then squeal as they settled onto the long concrete runway. There'd been no room to stow heavy, retractable, wheeled main landing gear in the XOV's underbelly. A clever engineer, maybe old Earl, had suggested reaching back into history and reviving the ski-like skids that had worked so well during the X-15 rocketplane program. With modern materials, those skids had served the XOV well, even during paved-runway landings.

Griffin slowly brought the spaceplane's nose down until the two-wheel nosegear touched, bounced, then settled back to the runway. There *had* been room for a retractable nose gear, which gave the XOV pilot a degree of directional control once all "feet" of the bird were in contact with the runway. He activated split-elevon speedbrakes and felt the tug of deceleration throw him forward against his seat and shoulder harness. Finally, the vehicle stopped. He quickly switched-off the auxiliary power unit, secured the cockpit, and waited until his ground crew gave a thumbs-up before opening the cockpit's canopy. Then the familiar wave of postflight weariness washed over him as adrenaline dissipated.

An hour later, Griffin had placed a "mission-completed" phone call to General Aster, conducted a detailed flight debriefing with his maintenance and control-center crews, and was ready for a shower and hot meal. But he wasn't finished just yet. He scanned the smiling, almost giddy men and women who crowded the small debriefing room. The one-star-general pilot thanked them for their backbreaking, night-and-day efforts to get the Blackstar system out of mothballs and into the sky so quickly.

Then he paused before adding, "We can take great pride in completing a very successful, demanding mission. Almost like old times, wasn't it?" A murmur of agreement rippled through the room. "But this isn't a development program, and that wasn't a flight test sortie. It was a space-combat mission. We restored a critical capability this nation had lost to an enemy attack. I suspect our next mission tasking will be even more demanding, so let's be ready. We can't relax, folks. This nation is still at war."

11

GAMBIT

"And, Paul. Tell us about that Army general—your Army liaison, I believe—we lost in that scrap with Hezbollah on the Israeli border this morning," SecDef Hurlburt said, pointedly putting the national security advisor on the spot. A hush fell on the teleconference line, broken by President Boyer's edgy follow-up.

"Paul . . . what the hell's this all about?"

Vandergrift blanched. *Doesn't anyone know what "top secret" means anymore?*

"It was one of *our* brigadiers, Paul," Hurlburt continued, a hint of fury riding on his words. "Of course, we know what happened out there, on the border. *All* of our intel satellites aren't on the fritz! What the hell were you thinking? What kind of 'leader' sends a general out where the enemy can get its hands on him? Especially one who just happens to know more about a damned Top Secret-Special Access Required info-ops system than any other human being in America!"

Hurlburt *really* wanted to strangle Vandergrift. Had they been in the same room, he absolutely, positively would have hit the smug weasel, and damn the consequences.

"What do you mean they *got* a general?" Boyer pressed. "I thought our boy was killed in that fire-fight."

So the president knew, Hurlburt mused silently, his gut tightening again. *Why even bother to have a secretary of defense?*

"Sir, we cannot allow these cowboy tactics to continue," Hurlburt said, straining to keep his infamous temper under control. "These terrorists keep beating the crap out of us because we keep acting like amateurs. I don't care what General Fisher was doing over there, but I guaran-damn-tee you it was *not* worth losing him—especially that way. Jeez! The bastards just dragged him away. A senior Army officer, for God's sake! Let's hope they kill him quickly, before those crazy damned Iranians get to him, because that's where he's going—to Iran. You can be double-damn sure 'bout that!"

Paul Vandergrift swallowed hard, trying to force from his mind's eye the image of Mike Fisher being hauled into Lebanon by crazed Hezbollah captors. A CIA intel flash, received just before the conference call began, had confirmed his worst fears: A business jet had taken off from southern Lebanon within an hour of Fisher's capture. A ramp worker had claimed that a man carried on board by known Hezbollah operatives was bloody and dirt-smeared, but was alive and conscious. Without question, the jet was headed for Tehran.

Vandergrift closed his eyes, took a deep breath, and took the only logical step open to him: start plotting his next move. *I hope that damned blockhead grunt doesn't spill his guts about SPECTRE before they kill him.* But hope was never a Vandergrift strategy. Bold action was required—and *now*.

16 APRIL/WHITE HOUSE, WASHINGTON, D.C.

Vandergrift scanned the EYES ONLY memo he'd printed moments earlier, trying to read its words from the perspective of President Pierce Boyer, his boss. The memo would take several days to garner the president's attention, thanks to escalating global tensions that were occupying Boyer's time these days. But it would serve its purpose: to prove that he, Boyer's trusted national security advisor, had notified the chief executive "in a timely manner" of covert, off-the-books operations "deemed critical to the national security of the United States." He knew the directives by heart. He had to, in order to routinely skirt them for his own purposes, yet adroitly deflect criticism and censure, *if* his exploits were ever exposed.

After all, this president had made it clear that he expected results, not blather, from his closest advisors—the same trusted people who had put him in the White House in the first place—and he really didn't need to know all the details. *Plausible deniability.* The insurance policy of every Washington politician, "PD" was the art of "not knowing" about an operation or activity that might prove embarrassing in the future. Yet, those were the same ops and activities that moved mountains in this town, the difference between *leaders* and gutless bureaucrats who simply maintained the status quo. History only remembered the *leaders*, and Paul Vandergrift was destined to be one of the greatest. He knew how the PD game was played, and had proven to be quite skilled at it.

The memo in his hands was Vandergrift's personal insurance policy, though. As soon as Charlotte Adkins's off-the-books political mission to Iran yielded what Vandergrift was telling himself would be a stunning breakthrough in U.S.–Iranian political relations, he wanted to make damned sure the president knew that *he*, not those weak-kneed eunuchs at the State Department, was responsible. And he expected to be rewarded accordingly, paid in the currency of Washington: real power, perhaps as the next secretary of state.

For a long moment, he reveled in the vision of Charlotte returning with a signed memorandum of understanding that would end decades of hostilities between the two nations. Iran, with its fiery, aggressive president, had become, over the previous five years or so, the political and spiritual leader of the world's millions of Muslims, subsuming the conflict between Shi'ite and Sunni into the broader conflict with the West's "crusaders." And now, in a single, brilliant stroke, Vandergrift would bring the Islamic faithful and the West's Judeo-Christian populations together in peace, thanks to the miracle of open trade. All it took was a little bending on the West's part, *and* an understanding that people everywhere yearned for one thing above all else: access to consumer goods! Hell, the route to peace in the Arab Mideast, Persia, and the Pacific Basin was painfully simple: *Build more malls!*

He had spoken to Adkins, his private emissary, earlier in the day, confirming that Charlotte had arrived in Tehran, and was scheduled to meet with Hassan Rafjani. Vandergrift had no idea who the man was, but he was clearly close to President Mahmoud Ahmadinejad, and the nation's Islamic Revolutionary Council, as well. Charlotte had sounded upbeat and supremely confident that her mission would be successful.

Charlotte Adkins's certainty was rooted in her overarching ambition to become Canada's prime minister, with the unfettered support of President Boyer, support that would be unleashed by the agreement she would coax from the Iranians. Her aggressive quest to be the nation's first female PM was an open secret in Ottawa, but none of her political enemies had any inkling that Adkins's strongest, most influential allies sat in Washington.

"Charlotte, dear, you sweet-talk that raghead nut case into opening Iran to trade with the U.S. and I will personally campaign for you in the Northwest Territories!" he'd laughed during their telecon. "Hell, I'll even go by dogsled to the faaaaar north to bag votes for you!"

"You're so full of shit. We both know what a liar you are, Paul," she'd snapped, but he could hear the smile behind the bite. "I can handle these carpet merchants, believe me. You just make damned sure this favor counts in Washington—and don't you forget who's doing *your* dirty work here! This isn't exactly Orlando and Disney World!" She'd described the rundown, drab appearance of Tehran, a far cry from what she'd expected. Despite its bluster on the world stage, Iran seemed to be a throwback, a déjà vu clone of the gray, colorless world of Communist-controlled Eastern Europe during the Cold War. Charlotte had admitted to feeling uncomfortable in Tehran, but was confident she could quickly wrap up her business and "get the hell back to civilization."

Vandergrift signed the memo and sealed it in a special envelope. His executive assistant would hand-carry it to the president's office, where, he suspected, it would rest in an untouched inbox for several days. Just long enough for Charlotte to finish her mission . . . and for *Sting Ray* to run its course.

17 APRIL/NATIONAL SECURITY AGENCY/FORT MEADE, MARYLAND/ SECURE VAULT T41

The room didn't exist. It was buried deep inside the forgotten confines of an agency that normally watched, listened, and waited. But not this time. Today, the agency would attack.

"Sadie, we've got five minutes to make the window. *Sting Ray*'s gotta swim. *Now!*" demanded the contract NSA computer specialist.

"I'm workin' it, Jaba. I'm workin' it. Damn Iranians made some config changes since we installed the trojan. Tricky. Bring me up on the parallel port once more," Sadie responded.

Sadie and Jaba were two of the agency's cyber-warriors, self-dubbed "The Doomers." They never used their real names, and their identities were known only to a few key people in the Information Assurance Division of NSA. There were approximately twelve Doomers in IAD at any one time, all of them counterculturists whose only religion was the hacker's code. They lived for the big game, hunting for weaknesses in their quarry's networks.

Although they rarely worked *for* anyone, many had joined IAD—they called it the "Foreign Legion"—following an FBI takedown of their group. The subsequent threat of significant prison time had a mind-focusing impact, as well. After their headline-making arrest, a National Security Agency liaison officer working at the FBI's headquarters recognized that the young, twenty-something Doomers' hacking skills were ideal for some very special projects. Born of a decades-long assault against U.S. military and intelligence computer systems by Chinese and Arab intelligence services, these special cyberwar projects had been decreed by a highly compartmented executive order that established a computer network attack or CNA capability within NSA. Although the Pentagon's Strategic Command had responsibility for CNA, it had focused its limited resources on network defense, unable to develop a costly, viable attack capability. In some minds, this, by default, cleared others to assume the "attack" mission.

Sadie tinkered with what appeared to be a collection of black boxes, flat liquid-crystal screens, and assorted diagnostic scopes, all draped with a disorderly array of cables and wiring. *A cyber junkyard,* thought Bill Toleen. Standing in one of the few clear spots available in the cramped and darkened room, Toleen was the NSA liaison officer who had argued for clemency, seeing more potential in the Doomers than most of his colleagues did. He had been on the Doomers' trail for more than two years as part of the FBI task force, getting a critical break only when one of their own finally compromised the group.

Their hacking prowess had reached alarming levels in May 2008, when they demonstrated their collective ability to hack into and take control of the Chicago Mercantile Exchange's trading filters, as well as the main operations center for California's giant utility, Pacific Gas and Electric. But the real shocker was their breach of the National Intelligence Directorate's community-wide counter-proliferation architecture.

The Doomers' caper had been highly synchronized—a simultaneous, multipronged strike. They'd left behind an organically developed antivirus that encompassed a blueprint of their actions—what could have been the consequences, had they pursued the hack further, as well as the characteristics of a self-learning antivirus left in the system. Motive? Nothing more than the thrill of a major intrusion. No harm intended other than a disruption of services and some severely bruised tech-egos. For whatever reason, the Doomers had showed a degree of benevolence in their hunt for the spectacular hack. Nowadays, vicious contemporaries in other nations wouldn't be so kind, if they were able to duplicate the stunt. Consequently,

Toleen had campaigned to bring the Doomers inside, to help NSA counter those really nasty cyber-snakes who could do serious damage to American systems and interests.

Hope to hell they can pull this off, Toleen continued to himself, *and slip that damn trojan into the Iranian air defense system command and control network. Jesus, these geeks are good!* The command network that the Doomers' software trojan worm had targeted was responsible for controlling all of Iran's strategic rocket and missile forces, particularly the ground-to-air network. *If they can get the worm to act like it's supposed to . . .*

The worm was a sophisticated virus, code-named *Sting Ray,* originally designed by one of the Doomers as a counter-virus, not an antivirus. The distinction was subtle and elegant. *Sting Ray* would not only detect and destroy a computer virus, but also retrace the virus' network path, then maliciously infect the perpetrator's computer system. The Doomers' work on the countervirus was timely. An "Eyes Only" executive order signed by the new president in March of 2009 called for the development of cyber-weapons to attack the seemingly impregnable security associated with any nation's nuclear or antisatellite missile forces.

There had been considerable debate about NSA's emerging role in cyber-warfare, but key members of the National Intelligence Directorate staff had little faith in the Defense Department's capability in such matters. Because NSA was the world's largest employer of mathematicians, and boasted the largest group of foreign-language analysts in the United States, many believed the agency was better-suited to conduct network attacks.

Toleen hoped his Doomers would become a catalyst for innovation in computer network savvy—and yes, daring—among the mix of scientists, engineers, and mathematicians working this critical project. He knew that many in the agency were appalled by having such riffraff anywhere near NSA's triple-fenced compound. But giving them access to the agency's most sensitive vaults and equipment? Ridiculous!

Toleen recalled one of the first attempts to socialize the Doomers with various division heads. The meeting had been as mutually antagonistic as any he'd ever witnessed: the buzz cuts versus the pony tails, milspeak versus technorap. But mostly it was the pension builders versus the anarchists, circling each other like rival gangs, neither approaching the large communal punch bowl at the same time. If the NSA engineers saw the hackers as predators, the hackers saw themselves as prey. Toleen chuckled, recalling the screaming and proclamations of disaster that would befall the agency if NSA established any formal relationship with the Doomers.

"Bill, you bring them aboard, and you'll have a mutiny on your hands," one had warned. After critics were reminded that the director had personally approved the Doomers' role, and that even the national security advisor and national intelligence director had been briefed and were supportive, the critics acquiesced, albeit under protest.

The classic Washington CYA dance, he thought, disgustedly. But it was the only way, he had told his bosses, that the United States could strike back at the command, control, and intelligence-gathering infrastructures of countries that posed a threat to, or had already cyber-attacked, the nation.

As Toleen watched Sadie and Jaba continue their seemingly undisciplined and highly haphazard approach, he reflected on the longest string of words he'd ever heard Sadie utter. During her get-acquainted meeting with NSA division heads, she'd said, "You people are boffo. Just read your manual! Lemme see . . . it says: *A computer network attack can alter the conventional war paradigm. Information intrusion can affect the social, cultural, economic, and military fabric of great societies. The potential application is boundless, since all elements of public and private sectors are dependent upon the power of technology. The great military mind of Carl von Clausewitz provided many of the principles for a modern information war as he reasoned that . . .* " She'd droned on, her mocking, bureaucrat-speak tone unmistakable.

But there was no mistaking the utter disbelief that

followed, and Sadie's follow-up had definitely endeared her to Toleen and the rest of this renegade bunch.

"Not shittin' in your shoes, Joe Square Pants, but is this *really* how you people write and talk around here?" Sadie had asked, waving the document. "No damn wonder your faces all look squishy, goin' around with your shorts knotted up in your asses. If that's how you all jibber-jabber with each other, I betcha that's how you think, too. No wonder we's always in the wrong damn dogpile! We don't even wanna know *your* music."

Toleen was jarred from his reverie. "Yo' Jaba! Got a match on our parallel port. I am *waaay* cool!" Sadie began to gyrate as if dancing to some beat of electronic pulses and primal responses totally alien to Toleen. High-fives and low-fives exchanged between Sadie and Jaba were a bit too much for Toleen's decorum. More importantly, there was the urgency of the mission at hand.

"Hey, you two, this ain't the Mardi Gras. How about we secure the attack port? Don't want *Sting Ray* swimmin' back upstream," he interjected.

"Got it," confirmed Jaba. "Now, Billo boy, you said we could peek at that directive for this gig. Why's my barbed fish bein' used to create such a mess in Rug Land?"

Toleen read the directive once more, an outreached arm shielding it from Jaba's playful grab-assing:

> *Multiple national sources indicate impending Iranian anti-satellite launch, possibly targeting unspecified U.S. assets. Recent loss of the International Space Station and critical U.S. satellites is ultimately attributed to Iranian hostilities. Requires immediate, aggressive action against Iranian strategic command and control networks. Computer network attacks authorized by NSA assets. Mission must be complete NLT 18 April.*

"You can probably figure it out," Toleen said. "We're working against a strategic missile and air defense network command and control, what we call a C2 system. So, let's just say we don't want any of *their* birds to fly, and leave it at that." The NSA professional flipped open his secure cell phone and reported. "*Sting Ray* is swimming." A sudden clamor of punk music made him wince and cover the phone's tiny microphone port.

Swimming it was. The *Sting Ray* program raced across fiber-optic cables and still-functioning satellite communication paths, transiting the same networks Iran's hackers had been using for years to probe the National Security Agency's computer system. But Iran had never cracked NSA's firewalls and isolated networks. Those same Iranian computer-hacking "soldiers of God," however, had violated one of the absolute no-no tenets of secure C2 networks: never, ever link them to the so-called World Wide Web or Internet.

Now, if *Sting Ray* worked, Iran's attack digi-bots would signal their human masters that they had successfully penetrated the NSA's inner, super-secret networks. The enemy software probes would appear to retrieve classified NSA information, when, in fact, the data mirrored only the Iranian program that had launched it, with a few tantalizing differences. Actually, those "data" were a malignant self-replicating virus, attaching itself surreptitiously to the very nodal receptors that also fed Iran's air defense and missile C2 networks.

Initially, *Sting Ray* would flood the nets with false data, fake passwords, phony asset sites, and nonexistent databank file names. Knowing how the robots should react, transmitting retrieval queries to all U.S.–programmed locations, the attack software would continue to replicate, sending the 'bots to new locations over and over again, all while duping Iran's hackers into believing they had successfully penetrated America's most secret computer files. Then *Sting Ray* would clandestinely feed the faulty information to the Revolutionary Guard's missile command, overwhelming it with patently false data, choking it to the point of insignificance.

But there were unknowns. Would Iran's network-defense software automatically detect virus-triggered overloads? Were Sadie and Jaba so skilled they could fool the Revolutionary Guard so quickly? Maybe. Maybe not. The Guard's missile command and control system might detect the *Sting Ray* attack, circle its digital wagons, and begin a methodical isolation and sector-protection process to counter the influx of corrupted, virulent data. But in a classic counter-countermeasures response, the *Sting Ray* virus was designed to strike yet again, using the adversary's own counterattack as a trigger mechanism. The virus would attach itself to Iran's new protective, isolation codes, then propagate a "sleep" command through the entire system.

Jaba had designed this particular hacker strategy as homage to his favorite episode of *Star Trek: The Next Generation*, "Locutus of Borg." In that classic, the android Data had sent a "sleep" command through Captain Picard into the Borg hive, shutting it down. *Ah,* Star Trek, Jaba thought, *the mother of us all.*

But all was not well along *Sting Ray*'s speed-of-light journey. NSA's band of scruffy cyber-warriors had been too haughty, too self-confident, succumbing to the heady wine of young nonconformism. The group's hackers had mistakenly believed they were smarter, cooler, faster, and more clever than Iran's older, university-trained computer and network scientists and software engineers. Unbeknownst to Sadie, Jaba, and Toleen, *Sting Ray* had been detected, trapped, and quarantined before it could unleash its havoc. A sophisticated countermeasure program had quickly neutered the "fish," while simultaneously sending the correct, expected indicators back to NSA, giving them the false impression that *Sting Ray* had succeeded.

Within hours, the source of NSA's attack had been identified, and the highest reaches of Iran's government notified. Hassan Rafjani was one of the first to be alerted. He demanded details, absorbed them quickly, and, unlike the less talented around him, recognized the tremendous gift he and Iran had been handed. He was unaware that Paul Vandergrift, the U.S. national security advisor, was the man behind *Sting Ray*, but Rafjani was quick to link the cyber-attack to the other Vandergrift initiatives: Charlotte Adkins and that

American general taken by Hezbollah at the Lebanon-Israel border.

The die is cast, Rafjani thought, reveling in his unexpected good fortune. A divine calm washed over his entire being. Now, God's war was inevitable—and its time had come.

However, Rafjani was faced with quickly convincing both the Ayatollah and President Ahmadinejad that an immediate strike against the infidels must be launched. In passionate terms, he carefully described for the president the full meaning of what the leaders' idiotic moderate advisors were trying to pass off as just the latest shot in a long-running, harmless cyberwar between the U.S. and Iran. He must convince them all. This would be his finest hour.

17 APRIL/NEAR MIDNIGHT/TEHRAN, IRAN

"We must not be fooled by the infidels' subtle treachery, my brothers!" he stressed. "America's insidious attack on our air-defense system is far more than a sophisticated jab in cyberspace. *No*! This must *not* be tolerated! It can *not* go unanswered! The Great Satan has attacked the heart of Persia, driving its blade of digital darkness deep into the very soul of our nation's electronic guardians. Think, brothers! Why would the NSA devils try to pollute the powerful electronic wall of Iran's air defenses? What would they have to gain from such a clumsy maneuver? We must think as the devils think! We must strike now before the blade of the infidel sinks any deeper."

The man code-named Dagger by CIA analysts—a shadowy, powerful figure they had declared "clearly psychotic and dangerously unstable"—glared at the Ayatollah, the president, and a small group of close advisors, his penetrating blue eyes flashing with hot fury. A fleck of white spittle at each corner of his thin lips underscored the fury of his words, a hint of the rabid dog he had become upon learning of the American cyber-attack.

An elderly mullah, the Ayatollah's most trusted political

counsel, vigorously shook his weathered, dark-turbaned head in denial. The mullah's watery eyes held Rafjani's, neither wavering nor blinking. The Islamic advisor spoke in unexpectedly deep, strong tones, openly challenging Dagger's assessment.

"Yours are the fiery words of careless youth, my son. You fail to foresee the future as it will be, a future that will only guarantee the destruction of Persia and all its people if your path to war is followed. On such minor provocation, you would waste the nuclear demons we have sacrificed so much of our nation's treasure to harness? Have you considered how the world would respond to such rash, unbalanced retaliation? Will our friends in Russia and China understand that a so-called cyber-attack on Persian air defenses warrants launching a nuclear attack on defenseless people in Europe or Israel? I think not."

Rafjani seethed silently, his jaw clenched tightly. His arm ached to unsheath the ripple-bladed stiletto hidden under his robe and sink it into the cowardly mullah's throat. But Rafjani restrained himself, drawing on practiced resolve and an uncanny ability to read the sense of those around him. *Your time will come, old fool!* he fumed to himself. Outwardly, he bowed respectfully, allowing the religious leader to continue his attenuated argument of tolerance and appeasement, senseless though it was.

"And you don't think, with all respect, that the Americans launched that worm against our defense systems because our Hezbollah brothers captured their general? Of course they did! The worm was revenge—typical, mindless American revenge! And it was the beginning of the invasion Satan has planned for years! Now we must show what their revenge has brought them! Just as we bloodied them four years ago. Only this time with Iran's full might."

"Launching a Shahab missile, armed with one of Persia's precious few nuclear warheads, at Israel or Europe, as you advocate, would be suicidal!" the mullah exclaimed, throwing both arms skyward. "*You* think about the ramifications, Rafjani! Think about the horror such a strike would unleash on Iran!

Europe and Israel would scream for overwhelming retaliation, giving American warlords the very excuse they thirst for, letting them activate their long-held military operations plans against our country. You think those American soldiers we know have been concealing themselves in our country for the past five years are only sightseers?

"The American aircraft carriers now cruising the Mediterranean would launch waves of Tomahawk cruise missiles and Hornets, each bird-of-prey armed with nuclear weapons. Even our arsenal of Russian undersea missiles is not enough to stop them. The Zionists probably would react even before the Satanists, firing their nuclear missiles at Tehran! Our people would die by the thousands, maybe millions! *No, no,* Rafjani! The actions you call for would most certainly destroy Iran!"

He turned to the large desk, clearly dismissing the snake Rafjani. "Mr. President, I implore you to abandon this insanity! Do not be seduced by my brother's plea for senseless revenge! The American cyber-attack does *not* warrant such an insane response!"

Rafjani again nodded respectfully, intentionally letting an uncomfortable silence settle over the room. A buzzing fly landed on the president's dark-topped desk, its sudden immobility accentuating the pregnant quiet. Then he exploded, seized by the moment, his body shaking as he drew himself to maximum height, fist raised as he exhorted the mullahs to prepare for all-out war. In his own mind, everything was so crystal clear. The American general, delivered first by Zionists, then by Hezbollah to his very doorstep! Then the Canadian harlot. They send a *woman* to do this work, to tempt us with the carrot while they wielded the stick, an even greater insult to all of us. And now the preposterous, easily countered computer worm. He, Hassan Rafjani, had been handed his God-given destiny, all within a matter of hours. The imams had shone a light upon his path. There was only one journey, now—a journey to the final battle foretold by the ancients.

"Enough of this nonsense! The Great Satan has attacked! Iran's fate hangs by a thread, the precious seconds that separate our people from destruction slipping away as we listen to the whining of cowards!" he thundered, shocking the small group. Nobody dared speak such blasphemy against a mullah, much less in the president's chambers!

"The American cyber-attack is *not* trivial, as you would have us believe, old man! You do not see the truth, the long blade of the infidel inches from our heart. Look closer and understand! The cyber 'worm' was very specifically targeted against our air defenses. It was intended to blind us, to cleverly emasculate the holy warriors who guard our borders and skies! Our radars and their powerful brains, the computers that sift through the ghosts and feints of electronic countermeaures, would have been neutralized, reduced to mindless, blind kittens, by Satan's 'worm.' The doors to our sanctuaries thrown open to the crusaders' flying bombs.

"Had the cyber-intrusion been successful, our brave soldiers and airmen would have been rendered defenseless, unaware and unable to deflect American and Zionist attacks on our homeland! *And therein is the truth we must seize and act upon!*" he railed, pacing furiously before the president, who remained seated behind a massive desk, his expression reflecting disbelief. The Ayatollah and other advisors stood, riveted by the extreme ranting of a being they thought they knew. All eyes tracked Rafjani, the men's expressions betraying discomfort and stark fear. *You are mine!* Dagger thought smugly. *Now for the coup de grâce . . .*

"The Prophet Himself placed His holy sword in *my* hands. The imams speak to me, demanding that His will be done! The American spy-general died by His holy hand, His holy sword! Now, even if I have to go forward alone, I will do so, for I am His arms, His legs, His hands!"

Several of the most fanatical mullahs were taken aback, exchanging glances, fear mirrored in dark eyes. Rafjani was approaching blasphemy! He believed that he, Rafjani, was a body used by God himself! The man was mad! But he was undeterred by the horrified stares of the old mullahs. *Let them fear*, he thought.

"May the fires of war consume the evil ones!" Rafjani railed, his eyes raised to the heavens, hands spread in supplication, the very vision of servitude to his God. "The voice of the imam, the Twelfth Imam, resonates inside my head. I see with His eyes!"

He stopped suddenly and leaned across the president's desk, the fingertips of both hands resting lightly on its polished surface. Now, in soft, yet clearly enunciated words, he spoke the truth he knew they all feared, but refused to acknowledge: "America and its lap dog, Israel, have sent their spies. Then they fired the first shot, even if it *was* a silent stream of digital ones and zeroes. Their missiles, bombers, and nuclear weapons are certain to follow. Your Excellency, the aggressors will believe their worm was successful, that our air defenses have been neutralized. Iran's faithful intelligence agents already are seeing disturbing movements by the Zionist air force and the Satanists' aircraft carriers. The serpents *are* preparing to strike! An attack on Iran *is* imminent! Be assured, Israel and the American Navy *will* send strike fighters to bomb our heartland, convinced they can easily slip through air defenses they *think* are disabled. We must act immediately, or . . ."

"*Nonsense!* You lie!" the old mullah interrupted, scowling fiercely as he yelled. "There are *no* bombers at Iran's doorstep! Only a fool would fire a Shahab on such flimsy, unverified suppositions!"

"No!" Rafjani screamed, his voice now manic with rage. "There!" He pointed to a corner of the room, his bony index finger quivering as it slowly raised, until it hung in space. Every eye in the room followed, now focused on the three-dimensional corner where ceiling and walls joined.

"The Mahdi watches us, at this very moment! In this hall, He is watching each of us, judging us. Are we pure enough to receive the spirit? I, Rafjani, His total and complete servant, am ready to lead, *to be the new imam! To fulfill His word! To meet my destiny in His service!*"

Rafjani spread his arms before the Grand Ayatollah, the religious head of state, and dropped to his knees. "The time of truth has come! I *am* the physical presence of the Twelfth Imam! Would you turn from Him, from the Twelfth Imam, on His day of judgment? I think *not!*" Rafjani roared, eyes flashing as he searched the souls of each awestruck, frightened mullah who flanked the Ayatollah.

It was madness, of course, and they all knew it. But their stunned silence only emboldened Dagger. His crazed eyes bored into each man in turn. Arms still spread, he finally turned his face to the sky and screamed, "I am the Mahdi!"

Rafjani stood, ignoring the Grand Ayatollah's stare, and leaned closer, ensuring only President Ahmadinejad could hear him. "Mr. President, what will you say to Iran's mothers, when they bring their dead children to you, as they did in Beirut, pleading, 'Why did you not stop the Zionists and crusaders? Why did you allow them to bomb our homes and kill my child?' How will you answer? Will you say, 'I failed to launch our Shahabs before they and our precious nuclear weapons were destroyed, buried in their underground shelters, thanks to Israeli and American bombs, because an old mullah "woman" convinced me there would be no attack'?

"Will you be a glorious Persian warrior, one worthy of leading God's chosen, Mr. President?" Rafjani hissed. "Or will you succumb to cowards who would stand by and watch Iran's destruction by the crusaders and their Zionist killers?"

Rafjani's dark features hovered before the president's face, which, he thought, was a picture of uncontrolled fear. *Ah, fear! Always it forces the weak to crumble,* Rafjani flashed. *Paint a fearful future, and the weak will capitulate to the strong, who will save them all from that future!* He was seized by the spirit of the Mahdi.

But what Rafjani did not see was President Ahmadinejad's eyes quietly squinting in assent. Nor could he read the president's mind.

The man is a master, Ahmadinejad thought, admiring the snake's ability to frighten these old mullahs into submission. *The old fools are completely under his power. God's will WILL be done now!*

Finally, the president stood and, with a flourish of conclusive bravado, delivered his decision: strike before the infidels could destroy Iran's nuclear weapons and missiles. "The Great Satan taught the world about preemptive strikes in

2003. Now the spineless dogs will suffer the singing sword of Persia's own *self-protective* preemptive strike!"

A brief consultation ensued, ultimately tempering the presidential decision with the practical political details of who and what would bear the brunt of Iran's history-making hammer blow. Not Israel, because the tiny nation would retaliate with the ferocity of a cornered animal fighting for its very survival, backed by America's might. Not Europe, because the Euros' behind-the-scenes, quiet moderation of America's hardliners was still a vital element of Iran's long-term gameplan. They could not afford to alienate the pliable Europeans.

But a devastating strike against an American facility in the region would be tolerated by Iran's neighbors, allies, and even the Euros. Why? Because most of the world's peoples silently believed that the American cowboys had become too powerful, too arrogant, too rich. Whatever injury the United States suffered at Iran's hands, America obviously deserved. Nobody would openly say as much, but it was true, the Ayatollah assured his president. The global response to Persia's Shahab launch would be noisy and indignant, but harmless. Iran could strike the Great Satan and survive.

As the gaunt, hard-eyed Rafjani walked rapidly back to his office, he carried the culmination of a life's work: signed orders to unleash the dogs of Muhammad's war, a conflagration that would end all wars. America's crusades against Muslims would finally be answered in one swift blow, and the era of the Twelfth Imam would be ushered in at last.

12

SHAHAB

The Shahab-4 missile leapt into the night air with a thunderous roar, trailing a multicolored palette of fire blasted from powerful rocket engines. A Pasdaran general officer of the Iranian Revolutionary Guard relayed a launch notification and early flight progress to the Supreme Council in Tehran, as well as to Rafjani and the few mullahs entrusted to witness the nation's courageous, historic counterstrike. Their holy missile was on its way.

The destiny of Iran took flight in a milk bottle–shaped rocket tipped with a 100-kiloton nuclear warhead. It marked the culmination of a clandestine nuclear weapons development and concurrent disinformation program over the previous ten years. They had outmaneuvered the Western infidels, lulling Europe, in particular, into inaction while secretly processing uranium into nuclear weapons. The West would now be stunned, and would cringe at this demonstration of Iranian might.

Close behind this launch, after the predictable American and Israeli response, the Revolutionary Guard would exercise its autonomous charge to wage war. Long-developed plans called for the Guard to focus its small inventory of devastating nuclear firepower against Israel, wiping it off the map, while the rest of the world watched in horror, petrified with fear. As for the Palestinians turned into charred remains by the nuclear strikes? Collateral but necessary damage. They would be the martyrs granted instant salvation. And when the cleansing firestorm ended, a new trilateral axis would emerge, a powerhouse comprising Russia, the Muslim world, and China. The centuries of Western rule would be over, a dark era consigned to history.

Rafjani had first proposed the cunning strategy that led to Iran's Revolutionary Guard infiltrating Iraq's Shi'ite party, unleashing a civil war with the Sunnis until all Iraq was consumed in chaos. First the British were forced out. Then the Americans, a humiliating repeat of Vietnam as the thousands of students and Shi'ite demonstrators stormed the Green Zone while American helicopters ferried out personnel and equipment.

When the Revolutionary Guard rolled in, it was welcomed by the Iraqi majority—millions of Shi'ites. The Russians called for peace under duress, grudgingly accepting the hard reality that Chechen rebels were controlled by Revolutionary Guard operatives. *It had been so clever, a masterful manipulation of the United Nations Security Council and the spineless peace-at-any-cost Europeans,* Rafjani assured himself. Under a UN-brokered agreement, Russia had dutifully processed uranium for Iran's supposedly peaceful use of nuclear energy. Naive global appeasers were mollified, while Iran rerouted higher-grade uranium for its well-concealed weapons program. They processed the weapons-grade uranium right under the upturned noses of smug Western diplomats.

The Twelfth Imam, indeed, was coming. The Shahab now climbing into space was his messenger, tracking across the general officer's screen, part of a control complex hidden deep in an underground bunker beneath the Zagros Mountains.

Onboard computers commanded the Shahab-4's guidance system to commence a roll-and-arcing maneuver, bringing the missile to a more westerly heading. In less than twelve minutes, a blinding flash would signal the newborn greatness of Iran's people. It was their trust in almighty God that had brought them to this confrontation with the Great Satan.

As he watched the retaliatory launch from afar, Iran's president reflected on how quickly the NSA's network-attack worm had been detected, successfully contained, and eradicated by brilliant Iranian software experts. He was just as amazed at how routinely they worked wonders within these enormously sophisticated systems. He was just as amazed at how inept the Americans had proved to be, as ineffective as children.

After receiving confirmation of the missile's launch, Supreme Leader Ali Khamenei addressed an assembly of mullahs in Tehran, reminding them of their future duties and continuing need to confront the Great Satan. He held court, using the occasion to deliver an incantation of Persia's greatness, underscoring the power of Islam. His vitriol and hatred for the West in general, and America and Israel in particular, intensified the clerics' fervor.

Two years after the breakup of Iraq, following the Americans' departure, today's ritual of self-accolades was particularly satisfying. Self-indulgent smiles greeted the Secret Intelligence Service chief as he boastfully recalled a U.S. plan in the late 1990s to curry favor with an Iraqi nationalist party leader then in exile. That exile's reams of false intelligence about Saddam Hussein's weapons of mass destruction program, augmented by a well-crafted Iranian deception campaign, had led to the West's conclusion that Saddam was an imminent, dangerous threat. All of it had played to Iran's overall strategic plan: convincing the United States to invade Iraq and bog down in the region.

The Americans were easily duped by the evidence, then further manipulated by Iran's SIS, which served to embolden the West's petulant march to democracy's siren song throughout the Middle East. The strategy was blessed when Americans initially disbanded the Iraqi army, a direct outcome of a daring plan developed by Iran's new heroes of the revolution.

Another master stroke was the Supreme Leader's quietly reaching out to the pacifist 2008 American presidential candidate, convincing him that, as a corollary to the regime's peaceful nuclear energy plans, it would cooperate with the West. Then, while the president filled the coffers of his lobbyist friends, the Revolutionary Guard conducted clandestine operations in Iraq following the upset American elections. These elements combined to trigger immediate withdrawals of U.S. troops from Iraq.

The ensuing Iraqi civil war had been short and bloody. Iran had played the role of peacemaker, gaining both territory and influence, in the process. Iran split Iraq into a toothless Kurdish state in the north—more a threat to the Turks than to others—with a tiny Sunni area around Baghdad, and a Shi'ite majority throughout the valuable oilfields, subservient to their Iranian overlords. The drive to a grander Islamic crescent, led by the purity of Iranian theocracy, was well under way.

While watching on a large display screen as their missile nosed over and streaked toward its destination, each mullah, in turn, proudly recalled other recent critical events. These were not confessions. Rather, they were orgiastic recitals of a diabolical frenzy.

They recalled their victory over the so-called Iranian moderates. How they had instigated student uprisings to expose the degree of foreign factors awash in Tehran. Elements of the Revolutionary Guard Security forces had seized massive caches of pornography, gay literature, anti-Islamist propaganda, and thousands of e-mails exchanged among Western governments, nongovernmental operative sources, Iranian students, and professors. Each of these "finds" had been publicly displayed, with e-mails and their authors' identities printed in daily newspapers and read aloud during television and radio broadcasts. Innocent Iranian professors, students, and officials of various government ministries found their names on those blasphemous and treasonous e-mails. Conveniently, all Western names of their so-called correspondents had been redacted.

No matter that the e-mails were all fabrications, that the pornography and gay paraphernalia had been planted by Security forces. They still inflamed Iran's citizens, even the moderates. As a capstone to the internal deception, several SIS-manufactured copies of "captured American, Israeli, and British war plans" that "proved" Iran was being targeted had been released to the Iranian press and global Muslim news media.

The mullahs had succeeded in generating outrage among tens of thousands of Iranians, aided, fortuitously, by the Danish cartoonists and their European monkeys who had blasphemed the name of the blessed Prophet by republishing the outrage. The decadent, stupid fools! The security forces couldn't have done a better job of inflaming an otherwise complacent Muslim population. Separately, inside Iran, anti-American zeal grew with each day's new revelation of an anti-Muslim conspiracy, all products of religious leaders' carefully crafted disinformation-and-deception campaign.

Show trials had soon begun in response to public clamor. Executions became daily occurrences, and a single, large uprising at the university had been met with overwhelming force, leaving thousands dead. The moderate voice of politics, religion, and culture in Iran had been stilled overnight. The mullahs had learned well, drawing from proven Stalinist disinformation techniques that had kept the Soviet Union functioning for more than seventy years: Those who control the media control the masses, and the bigger the lie, the easier it is to get away with.

There had been celebration and rejoicing as the mullahs savored their bold exporting of radical Islam. They were now accelerating Iran's ongoing, aggressive terrorist campaigns against the House of Saud, as well as the more secular governments of Lebanon, Jordan, Egypt, Turkey, and Thailand. Their destabilization programs were bearing fruit. Today, the Muslim world was improbably slipping further into an abyss of radicalism, dedicated to violence against infidels and collaborators alike.

Victory had been declared in the campaign to stifle a whiff of democracy that had briefly breezed through the Middle East in the mid-2000s, a campaign naively fostered by the West. Each mullah thanked Allah for his divine intervention in the 2008 U.S. presidential election. Its outcome had been a vital turning point for Iran and the mullahs' grand plan.

The Supreme Leader signaled that the session had ended by rising from his ornate chair. Ali Khamenei surveyed the mullahs and declared, "The world will know that America and the Zionists struck first, invading our missile-command system with their evil serpent worm and insulting the Prophet by sending a *female* emissary. The world will understand our sacred duty to defend Islam and our great people by striking swiftly, before fire rains down on Persia. Our Muslim brothers will share our outrage. And because the infidels of America and Zion are preparing yet another attack, even as we speak, we have begun the final battle."

The Shahab-4 was accelerating now, diving toward destiny.

18 APRIL/SBIRS SATELLITE CONTROL FACILITY/ BUCKLEY AFB, COLORADO

A Space-Based Infrared System spacecraft in geosynchronous orbit approximately 22,300 miles above the surface of the Earth detected the hot missile plume within seconds of launch. An Air Force controller watched, transfixed, as the satellite's staring sensors locked onto the missile body. While maintaining that lock, the sensors fed data to the mission control station at Buckley AFB, near Denver. A series of automated alerts appeared in the North American Aerospace Defense Command (NORAD) operations center, deep inside the granite of Cheyenne Mountain, seventy miles south, outside Colorado Springs.

18 APRIL/NORAD/CHEYENNE MOUNTAIN OPERATIONS CENTER

The staring sensor never wavered, tracking the Shahab-4 and transmitting a stream of data to ground stations. A precise course and projected warhead impact zone was soon displayed on a number of workstation screens. The NORAD command director initiated a missile-event teleconference, following protocols developed during the Cold War and rehearsed daily for decades. Key leaders alternately joined the conference, including the president of the United States and the Prime Minister of Canada.

The command director's most pressing assignment was to assess whether North America was under ballistic missile attack. If not, who was, and by whom? Tension was high, but months of drills by the NORAD watch team, each founded on years of tests and lessons learned from endless exercises, ensured a composed reporting of facts.

"North America is not, I repeat, *not* under attack," the director finally said. "Missile type is an Iranian Shahab-4; warhead impact zone is projected to be Aviano Air Base, Italy, home to Headquarters, 16th Air Force. Warhead yield unknown. No theater antiballistic missile ships or batteries available to counter."

"Italy? My God! Why Italy?" the U.S. president asked of no one in particular.

The command director continued his litany. "Warhead trajectory anomaly reported. Repeat, there is an anomaly. Updated impact is Adriatic Sea, southeast of Aviano. No additional launches detected."

"How long before impact?" asked the Canadian prime minister.

"Estimate one minute, thirty-seven, thirty-six, thirty-five seconds until impact. SBIRS has lost contact with warhead. Radar tracking data degrading."

"Is there evidence of an explosion? I mean . . ." from an anonymous voice.

"Command director. Impact in fifty seconds." Then a long pause. "Thirty-five seconds."

The teleconference line went silent. No muffled sounds. Nothing. Just stillness. An eerie quiet. More than a minute passed before the teleconference participants were jolted by another Command-Director announcement.

"Warhead impact past due. Negative NuDet! Repeat Negative NuDet reported." He paused, then added, "We show a release of radiation, but no nuclear detonation."

President Boyer was stunned, unable to fully grasp that an Iranian missile had been launched, but its weapon had failed to explode. He again spoke aimlessly. "Who were they targeting? I, uhhh . . . what is our required response to this outrage?"

The secretary of defense answered, trying to control the fury he felt. Hurlburt had predicted just such a scenario in 2007, during Boyer's campaign, but no one had seen a need to take actions that might have prevented it.

"Hurlburt here, Mr. President. There is no *required* response, but we do need to consider a variety of options, both offensive and defensive. I've ordered combatant commanders at Strategic Command and Central Command to prepare response options for your consideration."

"I understand. But I need to know: Why Italy?"

"If I may, Mr. President, why *Iran*?" Hurlburt said, an edge to his voice. "Our question should be: Why would Iran launch? They've always been purely reactive. We didn't do anything that would elicit such an action—"

"They couldn't have known about our surreptitious attempt to—"

Paul Vandergrift, the national security advisor, interrupted the president. "Sir, *Sting Ray* may not have worked as hoped, but I don't understand why the Iranians would shift to a ballistic missile attack when all indications showed them prepping an ASAT."

"Mr. President, may I suggest we take this conversation

off line *immediately*?" pleaded the national intelligence director.

"All right. NORAD, anything else to report?" asked the president.

"Admiral Brohmer, Commander NORAD/NORTHCOM, here, Mr. President. We have nothing further at this time, but we'll keep this missile-event conference line open for another thirty minutes with the other principals."

The prime minister of Canada registered his departure from the teleconference, as well, suggesting that his foreign minister would contact the U.S. secretary of state for immediate follow-up discussions.

Those remaining on the conference line were strangely quiet, mulling over the national security advisor's comments. Inside his office at the Pentagon, Hurlburt fumed. *A strategic plot afoot called "Sting Ray" and no one tells* me? *Someone's gonna pay for that.* At STRATCOM headquarters, an equally angry General Aster gripped his receiver more tightly, wondering silently, *What'n the hell is "Sting Ray"?*

18 APRIL/STRATCOM HEADQUARTERS/BATTLE STAFF WAR ROOM

The SecDef immediately called Aster on a separate secure line. The general was expecting it. Thanking Annie, his assistant, Aster drew a deep breath before he picked up the secure phone.

"Aster here."

"Can you tell me what in holy hell a damned *Sting Ray* is, General?" bellowed Hurlburt. "Don't feed me any B.S., either! I've had a bellyful of that, gettin' through that damned spaceplane business!"

"Mr. Secretary, I wish I knew. Our IO commander is on his way up here now. I suspect *Sting Ray* is some sort of info operations event, but—"

"Must be a *mighty* big event, General. The Iranians just tried to launch the first-ever nuclear strike since Nagasaki! Look, I meet with my staff and the chairman of the Joint

Chiefs in forty-five minutes. I want you to turn over every damned rock out there and tell me what the hell is going on!"

"Roger, sir." *How could this happen?* Aster thought. *How did we all of a sudden reach the brink of war with Iran? Christ almighty! Someone's running around half-cocked, and Hurlburt's got us pickin' up the pieces after the shit hits the fan.*

"Guess I don't have to tell you to activate your war plans, General. We'll be at ThreatCon Delta shortly."

"Understand, Mr. Secretary. We'll be developing a set of recommended options for you and the president."

Without warning, Aster's receiver went dead. No dial tone on secure phones; just dead air space between ear and receiver. *Too damned much dead air space emanating from Washington these days. Or maybe it's always like this. That town is filled with huge egos, always fightin' for position and power, yet no one's ever responsible for years of strategic malaise. Jesus, can't we anticipate some of this stuff?* He sighed and steeled himself for what he knew would come next. *Time to reengage Lee and his gamers.*

Aster shook his head, mumbling. As he turned the chair, his aide, Burns, confronted him abruptly. "With all due respect . . . dammit, sir, *we* attacked Iran! It was a computer network attack op, code-named *Sting Ray*. Target was Iran's strategic missile C2 network."

"Shit, Burner. Back up! Who is 'we'?"

"Sir, I've always been straight with you, but my info comes from a personal source who's protected my six forever. And knowing the president and SecDef are probably gonna have a meeting of the minds . . . well, I don't want my buddy to be a fall guy in this."

"Burner, stow that B.S.! This nation's security is at stake— or did you not pay attention during that missile-event conference? If and when heads roll, your buddy has nothing to worry about. I'll take the hits. But this is *no* time to play protect-my-agency's-turf! Nor am I in the mood to spend time convincing you about the propriety of your source's information!"

Aster looked Burns in the eye, and said, "We—you, me, and every other uniform in this command—are going to do everything we can to save the republic from all enemies, foreign and domestic. You got that? Now tell me, who conducted this IO attack against Iran's missile C2?"

Burns backed a step, looked around the STRATCOM commander's office, then tilted his head and whispered, "It's the NSA, sir. A special unit deep inside, using hackers they recruited after an FBI sting. Those renegades launched the attack, a computer worm. It obviously failed."

Burns recounted the *Sting Ray* event, adding that the national security advisor and NSA director knew, but nobody else, apparently. The entire op had bypassed the established info-operations chain of command. Aster seethed as he returned to his desk and picked up a secure line to the SecDef. Waiting for the call to go through, he turned to his aide again. "Get Admiral Lee and General Forester in here. And bring Androsin along, as well."

Aster clamped the phone to his ear, waiting. *So this is how World War III starts. Dead satellites, a restless public, scared-shitless politicians, and a stark Pentagon realization that we're vulnerable. And now our enemies know it. Man, now it's open season for miscalculations by fanatical or just misinformed leaders. The whole damned national security system is breaking down! All started by dead satellites! Hell, someone was crying about* that *wolf a long time ago, but nobody in D.C. answered. And now . . .*

19 APRIL/TEHRAN, IRAN

Rafjani glanced around the stone courtyard, darkening quickly as the last light of day disappeared behind ancient walls. The well-staged elements were in place. Several robe-clad, bearded mullahs stood in a group. They were God's witnesses—and his unquestioning validation, if necessary. An al Jazeera camera crew was set up, its thin-shouldered Qatari reporter obviously nervous. He couldn't stop pacing back and forth, a hand-held microphone tapping an open left palm incessantly. Finally, the Revolutionary Guards marched in, rifles on their shoulders. They halted, left-faced smartly, and uniformly snapped to parade rest, rifle butts on the ground, one arm extended, the other flattened against the small of their backs.

Rafjani turned and nodded, signaling an assistant. The Canadian woman, Charlotte Adkins, was pushed and dragged into the stone courtyard. She stumbled, but a dark-clad, bearded man yanked her upright, eliciting a grimace but no sound. Adkins locked eyes with Rafjani's, hers flashing with terror. His remained dead, cold. Hers continued to stare as she was brutally pushed to a spot near a pock-marked stone wall. Her hands were bound behind her, arms pulled back in such a way that her ample breasts protruded prominently. Her hair was tousled, a dishwater-blond strand hanging over one eye. Her silk blouse was torn, but revealed nothing.

Adkins straightened to full height and thrust her chin out, blowing the unruly strand of hair aside. Her lips quivered, Rafjani noted. *Good. The she-devil will enter hell in fear.* He turned his head again, this time meeting the wide-eyed, frightened stares of the al Jazeera reporter and his cameraman. Again, Rafjani nodded, then walked to a preplanned position before the TV camera.

"The Supreme Council of Iran has found this American infidel guilty of spying, and has sentenced her to death. God has spoken. His decree will be consummated," he said, speaking in Farsi. He stepped off-camera, then directed the operator to focus on Adkins.

She desperately tried to still an uncontrollable shaking that threatened to collapse her rubber legs. Biting her lip, she winced, then tasted the salty tang of fresh blood. That painful, sharp diversion steadied her racing mind for a moment. *Never look bad on camera.* The silly thought flashed across her consciousness from nowhere, so incongruous with the horror movie in which she found herself that she laughed. *Laughed!* That, too, struck her as incredibly hilarious.

Even Rafjani appeared to be stunned by her outburst, steadying her even more. Drawing a ragged, full breath, she shouted in English, "*Canadian*, you bastard! I am *not* American! *Canadian!* I am a simple messenger of peace! You are condemning an innocent, a servant of God who came to Iran in peace! May you burn in hell, you son of a bitch!" Anger, explosive, unbridled, all-consuming anger, extinguished fear. She spit a mouthful of blood at Rafjani, who recoiled. Without hesitation, he raised a palm, and waited, eyes fixed on the row of armed soldiers facing the prisoner, rifles raised.

Charlotte Adkins closed her eyes, anger evaporating as quickly as it had rolled over her. She fought tears of disbelief and frustration, lips again quivering. This time, biting didn't help. She felt, more than heard, the rifles bark. Blinding pain lasted but an instant, then darkness, blessed, peaceful darkness, as life slipped from her crumpled body.

19 APRIL/THE PENTAGON

The Secretary of Defense scanned an *Early Bird* compilation of national and world press clippings. He noted the expected outrage over Iran's attempt to attack Europe with a nuclear weapon, even though there was little doubt that the U.S. air base in Italy had been the missile's target. Two of the many press headlines were representative of the predominant Western message: IRANIAN NUKE AIMED AT ITALY FIZZLES AT SEA. IRANIANS LAUNCH BALLISTIC MISSILE; NUCLEAR WARHEAD FAILS TO DETONATE. As T. J. Hurlburt quickly scanned classified reports alongside the *Early Bird,* he was relieved to see that most intelligence analyses confirmed that even unfriendly regimes were backing the official U.S. position and those of most European governments.

But there were other headlines, he noted, from Islamist news outlets: AMERICA CONTINUES NEST OF LIES ABOUT IRAN. U.S. NAVY IS SOURCE OF ATOMIC PLOT. IRANIAN MISSILE LAUNCH IS ZIONIST AND AMERICAN HOAX. Al Jazeera, always represented by the faces of formerly well-respected American and British correspondents, carried the propaganda across the Third World.

The information campaign is already lost to the fundamentalists, grimaced Hurlburt, shaking his head in disgust. *And there will be consequences.*

He rose from his executive chair and stared at a photograph of his commander in chief, the president, and wondered, *Are you up for this, old friend? The Iranians are going to keep trying, and we'd better be prepared to take some pretty extreme measures. And I don't have many good options to give you. But there's gonna be people clamorin' for a nuke counterstrike, I know it. It's what I would do in your position. But I'm not. Our response must be precise, stealthy, and deniable. And absolutely, totally lethal.*

An aide came to the office's door. "General Aster on secure for you, sir."

He punched a speakerphone button. "Hello, Howard. How's your team doing?"

"They're working hard, Mr. Secretary. They're beginning to believe they live in this bunker."

"Yes, and that damn sure ain't gonna get better soon."

"Let me be frank, Mr. Secretary. I have unofficial word that *we*—specifically, the NSA—conducted a computer network attack against Iran's strategic C2 networks. Back-channel intel suggests those original CIA reports about the Iranians getting ready to launch an ASAT missile were wrong." Aster held his breath and winced, waiting for the SecDef's outburst.

"Godammit to hell! *What* is going on here?" Yelling for his aide, Hurlburt ordered, "Get me the national security advisor! *Immediately!* Howard, you keep a lid on this . . . this . . . Geez! This is a most *egregious* act! And the spook bastards never told me this attack was going down!"

Aster thought, *Seems more pissed about being kept outta the loop than about the attack itself.*

"And you're thinking the Iranians launched the nuke in retaliation?" Hurlburt asked.

"Yessir. It's entirely plausible. But our real concern here is: What's next?"

"Do we know their capability to launch again? Maybe that *Sting Ray* thing did some damage and that's why their nuke failed," said Hurlburt, hopefully.

"We don't know. The fact they were able to launch at all means that the IO worm didn't have its intended effect. I suspect the warhead falling short of its target and going low-order is associated more with our degraded GPS system, though. Iran's bright sparks just didn't account for the guidance variations due to GPS errors, which they obviously didn't know about. Can't speak for the nuke not going high-order, but the high-explosive package sure as hell worked. It scattered hot uranium all over. We've got a dirty piece of Adriatic Sea now."

"Right, Howard. But we can't assume the next one *won't* work, right?"

"Agreed, sir. And we'd better watch for other Iranian mischief. They're still paranoid about us, regardless. And, with all due respect, sir, regardless of their perception about *our* will to retaliate."

"You mean our president."

"Sir, if I may. We know the Iranians pulled off a stealthy, but strategic victory as far as the entire Iraq war was concerned, and they were behind the breakup of that country. They're still wary, however. They see us in every shadow, and their strategy is to hurt us so much we don't dare continue a campaign against them."

"Yep, I concur. They're flushed with their other successes over the past four to five years, and out of the blue, we try to bean 'em. Or is that *worm* 'em? They will *not* sit back and wait for us to respond, I guaran-damn-tee. And hell, they've been telling us to stay clear of their business for quite some time."

Hurlburt rushed on. "But I still don't get it. What in Allah's name made 'em prep for a launch to begin with? That damn sketchy intel we got from an on-scene source . . . what was that all about? I mean, we had a heads-up about this, didn't we? Or did something vital get lost in the static?"

"That's a question our gamers are wrestling with, Mr. Secretary. Admiral Lee's Red team is making some real sense, and it's raising eyebrows around here."

"Whaddya talking about?"

"That 'company' agent caught the Iranians preppin' something. Their intel analysts decided it was an antisatellite weapon of some sort. But I suspect the in-country agent really saw that nuke missile they just tossed at Italy. We'd have to scour the intel reports for his message, but I doubt we'd find much. Regardless, the mullahs didn't just decide to load a nuke in the last several days. What we're starting to think as we game this through is . . . well, we may be seeing a much longer-term strategy unfolding here, and I suspect Vandergrift simply handed them the smoking gun. That worm, or something else we don't know about, may have torched off their paranoia, which triggered the missile strike."

"Howard, don't dick with me, throwing charges like that around!" Hurlburt barked, exasperated by the dirty political shenanigans swirling around him. "The national security advisor's waiting on the other line. Get to the goddamn point."

"A 'tipping point,' sir. We may be at a tipping point. Our enemies develop multiple strategies, just as we do. Prepare for any contingency. They've spent years, even decades, thinking through the demise of America's influence. Well, what better time for someone to strike than now? The loss of our space systems has created an enormous window of vulnerability, and the bad guys know it. What we're gaming, sir, is that this could be *it*."

"Good God, Howard. Look, I'll address this at the NSC meeting. I want more of this solid thinking *and* some candidate responses from your team, ASAP!"

"One more thing, Mr. Secretary. It's unlikely the Iranians are the only ones thinking like this."

"We're on the same wavelength, Howard," the SecDef said as he stood, punching off Aster's line and selecting another blinker.

"Hurlburt here. That you, Paul?"

"Yes, T.J.," the national security advisor responded testily. "Can't this wait until the NSC meeting? We're in session in less than thirty minutes."

"No, Paul, it can't. I want to know just what'n tarnation this *Sting Ray* business is all about. And, Paul, don't bullshit me."

"*Sting Ray* is a computer network attack device, a very sophisticated digital worm, that was developed to preemptively disable Iranian strategic C2, their ballistic missile and ASAT systems, when deemed necessary. We—"

Hurlburt interrupted, his words tinged with barely contained anger. "I was never informed about this damned worm! You committed an act of war without NSC and Pentagon coordination, and without SecDef input? That's absolutely unconscionable, Paul!"

"We didn't go to *war*, T.J. A worm doesn't constitute an act of war. Besides, it was the president's decision to employ the worm, and to do so under conditions of utmost secrecy. You *do* understand our concern, given the number of leaks coming out of the Pentagon? Your pack of mongrels will do anything to sabotage this president."

Hurlburt paused for several seconds, trying desperately to keep his retort civil. Silence conveyed its own din. When he spoke, the voice was calm, yet commanding. "Paul, do not speak about people in this department in that manner *ever* again. You are absolutely delusional if you think a computer attack on a strategic air defense C2 system is not an act of war. And you and your White House moat dragons are *not* serving your president well, not at all. I intend to tell him so, too."

"That is your prerogative, but I would be careful about how you—"

Hurlburt snarled, "I do *not* need your dumbassed counsel,

Paul! I *will* speak to the president about the dire consequences your foolhardy worm mission has triggered, as well
as the inexplicable action of not telling *me,* the guy who's
responsible for the defense of this nation! Further, I will advise the president that he ensure Defense is the lead agency
to coordinate any planning and execution of all U.S. government responses to satellite outages and that Iranian missile
shot."

"That's the last thing we need, T.J.," Vandergrift said, his
tone intentionally patronizing. "Right now, the president is
more concerned with countering an Iranian disinformation
campaign, not to mention a sympathetic Muslim press and
skepticism in key power centers around the world. Besides,
there are diplomatic initiatives also under way behind the
scenes that can defuse the situation without bloodshed."

"Oh God! That disinformation campaign already has your
White House in a tail chase, and you're chasing the wrong
damn tail!" Hurlburt exclaimed. "If I hear another senator or
retired general on television spoutin' off about losing the media war, I'll . . . I'll take a dump on the White House lawn!"
Hurlburt was shouting. "Dammit, Paul! Screw the media
war! We've got the makings of a *nuclear* war, for God's sake!
And no one is paying attention to that little fart in the wind!"

The national security advisor responded with matching
rancor. "We'll see. And by the way, the president correctly
assigned the mission to NSA. I mean, uh, given the Neanderthals at STRAT, as evidenced by the sordid behavior ac-
corded my NSC rep, Preston Abbott!"

T. J. Hurlburt paused again, but thought better of expressing
further unkind thoughts. "Have a good meeting, Paul." He
slammed the phone into its cradle.

America's secretary of defense carefully weighed his options, then grabbed the receiver again, pressed a coded button
that provided near-instant access to the president, and with
deliberation and frankness spoke calmly. "Mr. President, I
am calling to report the malfeasance of Paul Vandergrift and
other NSC staffers. We are, at this moment, at war, sir, and, as

we speak, their recent courses of action defy reason. They are placing the very security of this nation in jeopardy."

"Go on," said the president stiffly, his usually formal tone now icy.

"You recall the Malcolm Gladwell book that came out ten years ago, the one that described how seemingly little things could gain momentum, reaching epidemic-like qualities? Then comes a defining moment, when suddenly you reach a 'tipping point.' You might recall that it was in your prep package before the global warming debates during the campaign. Mr. President, I fear we stand on the verge of such a moment. The Iranian nuke attack is just the first domino to tip. What if it falls?"

His only answer was a long, long silence. Boyer's strained voice finally responded. "T.J., do you have a TV on there?"

"No, not at the moment. Why?"

"CNN just aired a clip from al Jazeera. It . . . they . . . uh . . ." Boyer cleared his throat and tried again. "The damned Iranians just executed a woman by firing squad. On TV, for God's sake!"

"What woman?" Hurlburt was confused.

"Charlotte Adkins, the former Canadian ambassador." Boyer's voice was barely audible. "Some Iranian . . ." he paused, as if searching for the right expletive, ". . . bastard said she was an American spy."

Hurlburt gripped the receiver, his mind racing. What had Vandergrift said? *Diplomatic initiatives behind the scenes . . .* Suddenly, the missile launch and its dud nuke made sense.

"Mr. President, I think we'd better have a talk with Paul Vandergrift. That woman may have been another of his off-the-books ops." He let that sink in, saying nothing more.

Boyer groaned, mumbled something unintelligible, then hung up. Hurlburt slowly picked up a remote and switched on the wall-mounted flat-screen TV, selected a channel, muted the audio, then fast-scanned the preset, automatic digital recorder's memory. Seconds later, he watched, stone-faced, as a tall, beautiful woman screamed at the silent screen, then jerked backward and fell in a twisted heap.

19 APRIL/STRATCOM HEADQUARTERS/
WARGAMING OPERATIONS CENTER

Jill Bock felt the soreness radiate through her leg and back muscles. Fatigue was now a factor, and she knew it would soon affect the gamers' and support people's judgments. She looked up at Androsin as he returned from Aster's office, admiring his unflinching dedication and iron will. *Need to find a way to spell him,* she thought, which triggered another. *And need to help him sort through this stack of new reports. Gonna be hard to keep up with everything that's happening. Changing too damn fast.*

She stood to greet him, but his sullen look told her there was another rush of bad news in the offing. The colonel touched her arm, then hesitated. Over his shoulder, she saw Admiral Lee and General Forester enter. Both mirrored Androsin's obvious concern.

"They fired a worm at the Iranians, who then executed a former Canadian ambassador," Androsin said, speaking rapidly. "The worm hit Iran's strategic C2 network. NSA had a trojan in place, and *they* fired the friggin' worm! Just set it off. Didn't coordinate with General Aster, not even SecDef. Nobody! Jesus Christ, Jill, we're really in a mess now!"

"Hold on. Way too fast. Are you saying *we* started this . . . ?"

"Yeah. A preemptive network attack to deny a supposed antisat-missile launch. *Supposed* is the term of interest, 'cause the nuke they fired has gotta be what they were preppin', not an ASAT. Why the bastards killed that woman, I have no idea . . . yet."

"Cripes! So, what's next?" she asked.

"Duh," Androsin said, with uncharacteristic hostility. "That's *our* job to figure out, isn't it? Gotta game the Red possibilities, and develop a range of options. And we have to jump into it *now!*" Androsin strode purposely to the pit, Jill in tow, a pile of papers in hand.

"Colonel," she called, "I tried to get an intel cut, but the spooks slammed their green door, so we can forget any timely results from them. Probably arguing about the PowerPoint format."

"What are these, Jill?" he asked, accepting the reports.

"Well, there's an update about riots in Muslim population centers, worse than the aught-six Danish cartoon wars. The best sources appear to be some ham radio operators and tourists with communicator-cams feeding blogs on the Web. They're both outpacing the dino-newsies—and, of course, the intel weenies aren't bothering to review those nontraditional feeds," Jill continued. "BOYD's distilled most of 'em, then integrated that info with open-source news and raw intel. We collated everything with vulnerability protocols and added the potential political impacts."

Androsin scanned the bold headers. With a hint of anxiety, he declared, "We better get these to Lee and Forester immediately." He entered the pit, continuing to sift through the reports, selected a few, and had a Marine Corps technician display them on the room's big screens.

"Don't worry about format, Gunny. Just put 'em up so I can access by title." He turned to the amphitheater's weary audience.

"Ladies and gentlemen, we're ready to add new information to your process. And, if I may, I'd like to ask General Forester and Admiral Lee for an update on the Iranian-worm fiasco, and, if possible, on the diplomat's execution." Forester quickly recounted the gist of heated discussions involving *Sting Ray,* including the NSC and NSA decision-making process behind its launch—at least as he understood it, based on the sketchy information at his disposal.

"And I have nothing more about the execution. Our analysts are still trying to sort that one out. Hell of a brutal act . . ." the general said, shaking his head.

Then Lee stood, but remained at his station. "History is replete with miscalculations that spiral uncontrollably in the aftermath of unintended consequences." The admiral was up to his stentorian prep-school locution. He paused deliberately to allow a ripple of uncomfortable shifting to subside, portending the seriousness of his comments.

Lee made studied eye contact with several participants and continued. "The nuclear-armed ballistic missile the Iranians fired is only the beginning. Although that warhead did not go critical, there will be many chain reactions yet to come. They could involve additional nuclear launch attempts, or some other means of delivering an attack, while the entire Muslim world rises to excoriate the West and demand blood retribution."

"Admiral, excuse me. Are you implying we have intel about other Iranian nuke-launch possibilities?" a gamer asked, shock evident in his tone.

"No, we do not; not specifically, that is. Our Red team will scrub BOYD data and we will put forth the most likely Red actions. While the obvious focus is on Iran, we must also look to other opportunists who wish us harm. You know the usual suspects, but we must consider the means and other complicating factors associated with unintended consequences. Let me give you an example: Iran has long feared they were the next U.S. target, following our invasions of Afghanistan and Iraq, as well as our not-so-covert operations in Syria."

"Sure, Admiral. And those were justified actions. We've been over this ground," the gamer pressed.

Patronizing the interruption, Lee nodded. "To your point, sir. These were justified actions from *our* perspective. From Red's viewpoint, they were invasions. Couple that with our rhetoric earlier in this decade, and the Iranians fully believed that the United States intended to invade their country. So, they suckered us into an Iraqi campaign, ostensibly under their own perverted homeland-defense strategies. Not to be considered sympathetic to the enemy," Lee said, his tone dripping with sufficient acidic sarcasm to strip paint, "but remember the *Rocky* movies of the 1970s? Rocky's opponents used the best gym equipment money could buy. Rocky drank raw egg yolks and ran through the streets of Philadelphia with scores of screaming, barely clad urchins at his heels, urging him on. He won. In other words, the Iranians deployed the best home-grown capabilities at their disposal. And they succeeded.

They let our own weight and momentum draw us into that mess."

Lee turned and addressed Forester. "General, I strongly recommend the Blue team prepare a series of options to counter multiple Red actions. And include Red's counters to Green."

"Would you clarify that, Admiral?" Forester asked.

"As you know, the Green team comprises the interests of our allies and neutral entities. Commodore Barker of the UK can help us understand allied responses, so we can better integrate our diplomatic coordination in near real time through State's liaison officers here. Very important, however, will be the Israeli response. Now that they've experienced a 100-kiloton nuclear flyby—a 'special delivery,' so to speak—from their archenemy, you can only imagine what *they* must be planning right now. For example, if we think they didn't detect that launch, prep their own antimissile defenses, and ready a nuclear response, we're functioning in an alternate universe. Consequently, we have to not only prepare for an Israeli gambit, but also develop some pol-mil options for the president, should the Israelis preemptively jump in and escalate this nuclear bar fight."

Lee strode toward the podium. "One other point. The Red team will also assume a role of the Muslim populace and examine their eager defense of Islam. I am speaking of elements well beyond the remnants of al Qaeda and their affiliates."

Androsin walked to the admiral's side, caught his attention, and whispered to him.

Lee nodded and turned to look at the overhead screens. One displayed streaming video of American embassies and Israeli interests being burned in the capital cities of nations with predominately Muslim populations. Scenes were punctuated by shocking images of hostages being dragged into the streets and shot or bludgeoned to death, then bodies being burned. Another screen portrayed real-time cell-phone dialogue, close-captioned as a scroll of comments: "They're killing all Americans and Westerners! They're coming for us! There's no way out! Please help! God save us . . ." Their shrill pleas and fear were evident, even as silent, flowing text across the display.

Al Jazeera was fueling the media frenzy throughout the Muslim world, whipping up viewers with a single word: *jihad*. Islamists were using every media and propaganda outlet at their disposal to exhort the masses, calling for retaliation and citing the U.S. worm attack on Iran. They dismissed as lies any claims that a nuclear weapon–tipped missile had been launched by Iran. As Iran's effective disinformation campaign unfolded, Islamist leaders railed against an assumed imminent invasion of the nation by the United States and Israel, claiming repeatedly that the two nations intended to destroy all Muslim holy sites in the Mideast.

Only passing mention was made of "an American spy being executed" by the Iranian Guards. Even that horror was twisted, presented as an appropriate response to yet another affront to Islam.

Apocryphal images seared the center's large display screens, stunning every wargame participant. Lee's attention shifted from the gut-wrenching images to the task at hand, his gaze taking in the wide-eyed team members across the amphitheater. *It will be hard to sort these out,* he thought. *Mob frenzy is driving leaders' decisions. Or is it the zeal of leaders driving the media and the mobs? Either way, this is serious. It's spreading to the Muslim ghettos of Europe, too. Soon, our own citizens will become incensed over the Adkins execution, their anger fueled by paranoia about sleeper cells here in America. An army of suicide bombers, both real and imagined, is waiting to be unleashed, while every local sheriff is on the lookout for anyone he decides is suspicious-looking. Christ!*

Lee reflected on Europe prior to the First World War and President Wilson's subsequent attempts to get the United States involved. The *Lusitania* had been a deliberate sitting duck. The world had been divided into armed camps a century ago, each side waiting for just the right spark to set war into motion, each camp frothing at the mouth like a snarling dog, anxious for an opportunity to send its armies into battle. It had been the subject of his Master's thesis at Princeton's Woodrow Wilson School of Government. *What fools they had been, those nineteenth-century Hapsburg behemoths, desperate to unleash the latest technologies of death upon their adversaries.* Europe would suffer for generations, as would the entire world, thereafter. *Have we learned nothing? Are we at the same place again, a hundred years later?*

Lee's mind wandered to his children and grandchildren, ever so briefly, then returned to the wargame's key issues. *We have to stay ahead of events. I wonder what our friends in China are thinking about all of this at this very moment. They're focused on disturbances in their western provinces, debating Japan's intentions and nervous about the crazy North Koreans going over the edge. And Taiwan. Always Taiwan.*

A growing murmur circulated throughout the wargaming center as participants grasped the full impact of what they were witnessing. The increasing gravity of a deteriorating global situation was readily apparent, underscoring the importance of their task.

Androsin stepped back into the pit and announced, "All right. Let's resume with a Blue-versus-Red move and counter-move session with the following objectives: One, avoid a wider war; two, deter and/or destroy Iran's capability to launch nuclear weapons; three, examine immediate space-control options to protect U.S. and gray—meaning commercial—space assets; and four, develop, in parallel, diplomatic initiatives aimed at deterring any additional aggression against the United States and its interests. Be prepared to brief General Aster on preliminary courses of action in three hours."

Turning to Jill, he half smiled, attempting to lighten the mood, and whispered, "Whaddya say we grab a beer?"

Matching Androsin's weak gesture of peacemaking, she

replied in kind. "Sure. Why don't we slip down to Cancún for it? Nothin' much happening here!" She glanced up at the latest images before leaving the center. She fought back tears as another body writhed in flames, this one bound in an American flag.

13

A Prophetic Soul

Military action is important to the nation. It is the ground of death and life, the path of survival and destruction, so it is imperative to examine it.

Sun Tzu

Major Clint "Mannix" Gleason nursed his B-1B Lancer bomber to 28,500 feet, pushed its nose down slightly, and waited for the airspeed to build. *Just a little more, baby,* he muttered to himself. Every foot of altitude and knot of speed mattered.

"Fifteen miles from the drop point, Mannix," Captain Lance "Grinch" McCartor, the offensive-systems operator, called via intercom. He and his defensive-systems counterpart were out of view, seated in a separate compartment a few feet aft of the two pilots. "Package is hot. Ready to open doors on your call." Gleason understood; the system they were about to drop was powered up.

"Stand by. We can get a little more out of this ol' girl, and I don't want the bay doors' drag just yet," he responded. Airspeed numbers on his flight-data display crept higher, then steadied. The bomber's four F101-GE-102 engines were at maximum power, each straining to deliver all the thrust both pilots' throttles commanded. "Deadwood City," the bomb wing's best B-1B, at least by Gleason's reasoning, was giving all she had.

"Okay, Grinch. Prepare to launch," Gleason said, glancing sideways to make sure his copilot was ready. A thumbs-up assured Gleason, the aircraft commander, that all prelaunch flight conditions had been double-checked. Sure, they'd like to be higher and faster, but with a heavy payload and enough fuel to get back to Lajes Field in the Azores Islands off the coast of Portugal, this setup would have to do. When it was heavy, the old "Bone"—pilot talk for "B-One"—never did like to cruise at high altitude.

McCartor opened the center bomb bay's doors, then carefully rechecked the data stream being sent from a cruise missile hanging on a rack in the bay. He raised a hinged red metal guard on the panel of controls and displays that rose vertically above his small worktable, and placed his left thumb under the now-exposed silver toggle switch. "Everything's good here. Drop in . . . three, two, one. Launch!"

The crew felt a solid thump as a modified AGM-86C Conventional Air-Launched Cruise Missile, or CALCM, was ejected from the center weapons bay. A moment later, scissor-like wings extended from the CALCM's body, locking into place. A compact Williams F107-WR-101 turbofan engine ignited, generating sufficient thrust to quickly propel the gray weapon ahead of the B-1B. Seconds later, the long, slim vehicle started climbing, pitching steadily upward until its nose was pointed well above the horizon.

"Telemetry solid. Bird's on profile; systems nominal," Mc-Cartor droned impassively, scanning a jumble of numbers and text flowing across a screen on the laptop computer screen anchored to his worktable. He hoped the same encrypted data stream transmitted from the CALCM to his bomber—plus his verbal comments—were being relayed to a commercial satellite transponder miles above, then down to a ground-communications network that fed workstations monitored by dozens of nail-biting engineers, managers, and USAF officers. At least, that's how it was supposed to work.

An Instant Message window popped onto his computer's display, confirming the payload operations team at Johns Hopkins University's Applied Physics Laboratory in Maryland was indeed receiving data. McCartor breathed a quick sigh of relief. If the APL operators were getting data, then the CALCM's payload had a good chance of reaching its destination and delivering as advertised.

The CALCM climbed at an 83-degree angle, almost straight up, carrying a highly unusual payload. Normally, the cruise missile would be armed with a powerful explosive warhead and would be diving toward the ground, its GPS guidance system directing it toward a target hundreds of miles away. Instead, this modified CALCM was serving as a cheap, on-demand booster, carrying a novel airship to the fringes of space. Gleason, McCartor, and even their wing commander had exchanged looks of doubt, each mentally raising and waving the proverbial "bullshit flag" when Johns Hopkins managers had first briefed them about the HARVe-2 platform.

There's absolutely no way you can cram a damned blimp, electronics platform, and two electric-powered motors and propellers into a twenty-one-foot-long AGM-86C missile body! Gleason had thought at the time. Logic said there simply wasn't enough room in the missile's body to carry anything that big, even when compacted into an unbelievably small package. But the ingenious APL boffins had done it *and* proved it would work quite well through several demonstration flights in 2008.

During those prototype demos, the High-Altitude Reconnaissance Vehicle (HARVe) had been tested as a communications-relay and on-demand tactical recce platform dedicated to serving regional combat commanders' needs. In essence, HARVe was a poor man's on-demand satellite, albeit flying at a much lower altitude. But it stayed in one place, was always available, and could be moved, if necessary. And it was one hell of a lot cheaper than a full-blown spacecraft, the B-1B pilot noted, turning to his return-to-base heading.

A second-generation HARVe system had been funded and was being prepared for another round of testing, scheduled to begin in mid-2010. But, when the U.S. started losing precious GPS satellites to maser attacks, smart APL folks stepped forward with a novel idea: use HARVe as an airborne GPS-related wide-area augmentation system (WAAS) to compensate for inaccurate signals. Ground-based versions of WAAS were already in service across the U.S., transmitting error-correction signals that substantially improved the accuracy of GPS navigation systems on aircraft. With some clever changes, the same thing could be done at the edge of space, they suggested.

From its perch above 100,000 feet, the HARVe airship's electronics package would beam error-correction signals throughout a broad geographical area, compensating for the inaccurate navigation and timing signals being spewed out by the degraded GPS satellites that had suffered maser attacks. It was a wild idea, but a quick engineering demo aboard a high-flying testbed aircraft had proved the concept would work. And it was a rapid way to restore some of America's crippled space infrastructure, at least temporarily, for one area of the globe.

Within days, APL had reconfigured a second-generation HARVe system, adding a GPS-signal-compensation package, and crammed it into an AGM-86C cruise missile one of the Air Force laboratories had modified. The missile had been loaded on "Deadwood City" and flown to this spot over the Atlantic in record time. Now, the missile was climbing into that no man's region Air Force Space Command called near-space, preparing to dump its precious payload.

"Through 100,000, nosing over," McCartor said. "Fuel's almost expended." He watched the fuel numbers tick down, expecting the CALCM's engine to flame out at any moment. There! A sudden drop in engine RPM signaled the expected loss of thrust. "Okay, Control; it's all yours." Switching to the B-1B's intercom, he exclaimed, "We did it, guys. Yeee-haaa!"

From the APL control room half a world away, a tense-voiced test engineer replied, "Control confirms. Stand by for separation." The CALCM was programmed to automatically start the HARVe deployment sequence when the right combination of speed, altitude, acceleration, and other parameters had aligned. McCartor's limited-purpose computer wouldn't display or let the officer affect that sequence. That was APL's ballgame.

Miles above the B-1B that was now banking and descending toward a landing in the Azores, the AGM-86C started coming apart as pyrotechnic releases were automatically fired by onboard microprocessors. Flawlessly, the missile's nose fell away, taking a GPS guidance package with it. The tightly packed HARVe platform unfolded, a gossamer, high-tech butterfly emerging from its war-machine cocoon. A tapered, bullet-shaped fabric balloon expanded rapidly, pressurized by small-diameter, high-pressure-helium bottles. Within seconds, the now-rigid structure stabilized.

Two small engine pods snapped into position, suspended from angled struts anchored in the balloon's flimsy aft structure, and a couple of two-blade propellers unfolded. APL had borrowed proven prop-stowage technology and mechanisms from the motorized glider community, believing there was no need to reinvent a system that had a history of reliable airborne performance. Those props started turning slowly, powered by highly efficient electric motors that drew sustenance from small, high energy-density battery packs and a blanket of dark-colored solar cells covering the upper surface of HARVe's blimp-like airship.

A platform slung horizontally under the blimp's tail and between its engines held the sophisticated electronics package that, if it worked correctly, would give U.S. troops throughout the Mideast and parts of Central Asia—Central Command's Area of Responsibility—accurate GPS capabilities again.

A command transmitted from APL's Maryland control center activated HARVe's onboard electronics package. Its error-compensation subsystem immediately started looking for downlinked GPS signals. At the speed of technomagic, selected GPS signals were devoured, compared with pre-programmed data in onboard microprocessors and sorted. Then error-correcting beams of data were transmitted from HARVe's Earth-aimed antennas.

20 APRIL/CENTRAL COMMAND HEADQUARTERS/QATAR

"All right! Yes! Wooo-hoo!" whooped a young Applied Physics Lab engineer, pumping a fist into the air, ecstatic. The commotion drew the attention of a small knot of Army, Air Force, and Navy officers standing nearby.

"General, you're in business!" the APL engineer said, pointing to a set of colored figures on a large-format laptop computer screen as the officers approached.

"Okay. But what the hell am I looking at?" growled a weathered Army one-star, peeved that the young geek expected senior combat officers to decipher techno-gibberish.

"Like . . . you have accurate GPS nav and timing data throughout the theater now!" the engineer enthused, his face beaming. A glance at the general's dark eyes and scowl erased the younger man's grin. "Uh . . . well . . . this column shows data coming down from this satellite here, see? That's a degraded GPS bird, one that'd been zapped by that maser weapon in . . . wherever-stan. So it's spitting crap . . . er, I mean bad nav data. A receiver on the ground . . . say, a soldier with a handheld GPS unit in the Mideast, would be getting bad timing and position data from the degraded satellite. He thinks he's somewhere other than where he really is, maybe as much as a few miles off. Those damaged satellites' onboard systems have really drifted, because the Air Force can't get updates—error-correction data—up to them. Haven't for days, so they're really going to hell," the engineer explained in a rush, trying to forestall another bark from the general.

"But this new gizmo, that near-space thing the B-1 just put up, was supposed to correct the problem, wasn't it?" the general asked. Crossed arms and furrowed brow, accentuated by a high and tight, very close-cropped haircut, said he still wasn't tuned in. "Just get to the point, son. Do I have good GPS or not?" he grumped.

"Damned straight, General!" The engineer's enthusiasm wouldn't be stifled. "The system on HARVe, the near-space platform that's . . . let's see . . . now at an altitude of 102,400 feet, slowly moving toward the Mediterranean under its own power. Approaching Gibraltar, roughly," he said, tapping a few computer keys. "That system's programmed to look at data from specific GPS satellites, the birds we told it were degraded and couldn't be trusted. Then it looks at the good satellites now in view and compares the position and timing data from both the good birds and the crap birds. An onboard algorithm does its magic, decides exactly where the HARVe platform is in the sky, and sends out signals that tell only certain GPS receivers down here, on the ground or in the air, what corrections to apply. And, like . . . bam! You know where you are again. And it's accurate!"

The general nodded slowly, a half grin starting to show. *The silver-bullet HARVe must be working,* he decided.

"Good. You guys did a hell of a job with this near-space fix. I'm impressed," the officer said, giving the engineer a thumbs-up.

A few more minutes of discussion among the engineer and other officers convinced the general that, indeed, his ground, sea, and air troops were assured of good GPS information. At least those in the large swath of Earth that could be reached from HARVe's high-altitude perch were, for now. But the Iranians, Mideast terrorists, and any other unfriendly GPS users would still be getting inaccurate GPS data. The bad guys didn't have access to the military-only P-code GPS signals, and they didn't know which GPS satellites passing overhead should be ignored. Hopefully they didn't, anyway . . .

The general started to leave, then turned and called, "Hey! How long can this HARVe stay up there? Am I gonna lose those GPS corrections after . . . what? A few weeks?" he asked.

"Not for quite a while, General," the APL engineer replied. "HARVe should be up there for several months. It's high enough to stay above thunderstorms and atmospheric turbulence, and it won't be running out of fuel. As long as the sun keeps shining, and the balloon doesn't leak and the 'tronics keep cookin', it'll keep pumping out good data that corrects GPS signals throughout your region."

The general waved his thanks and headed for a sparse office down the hall. He finally had some good news for his boss, and the STRATCOM chief, as well.

20 APRIL/AIR FORCE SPACE COMMAND HEADQUARTERS/ COLORADO SPRINGS

"That's right; we're restoring some space capabilities as fast as we can, Howard," Buzz Sawyer assured Aster via phone. "You know about Speed's spaceplane mission, of course. That gave us some additional imaging assets that're helping us fill the gap left when that Hunter took a hit. And we're recycling Blackstar as fast as we can. It oughta be ready to go again in a few days."

"We're gonna need it for more than just hauling stuff into space, Buzz," Aster noted, his ominous tone not lost on the other general. "But go on. You've popped a few other things into orbit, I hear."

"Rog. A few of the contractors came through, and we were able to get some limited-capability comsats—basically microsatellites—into LEO. We got 'em up there with a Pegasus air-launched rocket and one of the new Falcon boosters," Sawyer continued. "The comsats are really just gap-fillers for some of the Excaliburs we lost, but we now have more high-bandwidth comm over the Pacific. The Navy can at least talk to its task groups out there. Not perfect, but it's more than we had a few days ago."

"Great! Keep pushing those contractors—and our lab guys. They've all risen to the occasion, it appears." Aster had been pleasantly surprised to see how the aerospace community had pulled together, coming up with quick fixes that drew on disparate skills and proprietary hardware and software. At least the aero and defense folks had quickly recognized the dire consequences of a degraded space infrastructure, even if still-posturing and harrumphing politicians in Washington hadn't.

"Yeah, definitely. And I'm sure you were briefed about APL's HARVe near-space platform," Sawyer said. "So far, it's working like a champ. It's been moved and is station-keeping over the Med now. We couldn't launch it over the Mediterranean, where we wanted to, because the Euros were fretting that Iran, and even Israel, might get excited about a cruise missile being launched on their doorstep."

"Yep. Understand. Hard to believe the near-space platform is performing so well, though," Aster mused. "That concept's been diddled around with for years, and there're several types still in development, right?"

"Well . . . frankly, I'm holding my breath, Howard. HARVe *is* still developmental, and has almost no redundancy built in. Even the APL guys can't tell us how their electronics package will survive the severe temperatures and long exposure to cosmic radiation above 100,000 feet. But, for now, it's performing. CENTCOM's about to wet their pants over what it's restored for 'em!"

Aster chuckled. "We needed a touchdown, Buzz. Give your people and the APL folks an attaboy. They've damn sure earned it!" He signed off, cradled the phone and headed for the door, punching the air in an uncharacteristic display of enthusiasm. *About time the good guys scored!* the STRAT-COM commander thought, wearing a tight-lipped smile as he breezed past Annie's desk.

20 APRIL/STRATCOM WARGAMING OPERATIONS CENTER

Space system updates bein' fed to ya', my man. Jill always spoke to BOYD in human terms. It kept her mind on the task at hand, and helped provide context to every aspect of the gaming process. *Chinese and Iranian offensive counter-space capabilities updated; maser capabilities updated, meaning the damned maser is out of commission, finally. You're hummin' now,* thought Jill, savoring a sense of admiration for the people who had designed BOYD's architecture.

The powerful BOYD computer-based tool was receiving multiple space system reports, while automatically cross-checking orbit-analysis parameters from the Space Control Center and SIVTRIX. All updates were automatically dumped to a current game registry. *Okay, BOYDster, do your magic.* The space system info had never been more critical than right now. *Registry checks good; no BOYD anomalies. SIVTRIX displays cross-checked. Ouch! SIV's showin' some big-league holes in space coverage, and BOYD's painting an ugly picture of remaining vulnerabilities. If some other bad actor wants to jump into the fray, our space resources are hangin' out there; just limpin' ducks!*

"Ma'am, game registry update complete. BOYD's in analysis mode," called one of the technicians.

"Roger. Check that registry access is verified for all gamers' stations."

Jill flashed her best game-face smile, grabbed the BOYD-status printouts, and angled toward Androsin, who appeared to be in casual discussion with Admiral Lee.

"Beggin' the admiral's pardon," she interjected cautiously. Lee nodded politely. "Jim, BOYD's been updated with the latest space system status and capabilities data. We can advise the gamers anytime."

"Concur," responded Androsin crisply. "Let's dedicate one of the overhead screens for the 3D Earth-coverage depiction model, then get going. Agree, sir?"

"Indeed, Colonel," said Lee, always proper. "I'm anxious to get started on this move." The retired admiral returned to his workstation and, using a computer-screen stylus to activate an icon, sent BOYD's updated space system status data to all of his Red team players. Each member saw a new array of graphics and data sets that highlighted the locations and movements of Blue and Gray space systems. Via pull-down menus, a Red-team player could view Blue's current operational picture, such as communications bandwidth and links available, intelligence coverage, and other capabilities of active space systems.

Lee selected the aptly named OPPORTUNITY menu, having decided to focus Red on "space control" or "counterspace" issues. Those monikers had caused undue hand-wringing among allies and adversaries alike over the last two decades, but they were really only new terms for basic military tenets: protect your own space resources, and deny the enemy the use of his or others' space capabilities. The "how" of doing both, though, was among the most tightly protected information in the U.S. milspace arena.

Using the Red-only intercom channel, Lee directed his embedded space team. "I want the anti-Blue cell to use OPPORTUNITY for an in-depth assessment of Blue's current space vulnerabilities. Look at three space control areas, as a minimum, and others as you are able. First, explore space control actions that Red might resort to, with emphasis on China and Iran. Focus on actions that could immediately disrupt or deter Blue's offensive and defensive military operations. Second, determine space control actions that could strategically inhibit Blue's national leadership. Third, explore space control actions having the greatest impact on the social fabric of Blue and its allies."

Lee paused, reading an Instant Message input from a Red team member. *Mmmm, good idea.* He continued. "And be sure to integrate those space control findings with Red's diplomatic and military cells. Then develop your space control options. One thing we don't want: to behave as Blue might when confronted with tough issues like these."

One of the Red team members spoke up via intercom. "Admiral, are you saying you *don't* want us to work the tough issues through the chain of command, like Blue always does? If we integrate this info, and develop cohesive options for the highest levels of authority, won't we miss a golden opportunity to blame each other for lack of coordination?"

The sarcasm and burst of laughter it triggered broke a thick tension that had developed on the "Red Net." Lee smiled, but grimaced inwardly. Those comments, though humorous, were all too true of the United States, he thought. *This wargame* must *pull all these issues together, including the Washington players. This is not the time for stovepipes and turf wars. Right now, we don't have a good picture of what's happening, nor a good feel for the ramifications of retaliation. And there clearly is a growing public clamor for a response to the horrific attack on the space station. We better make sure it's the right response.*

Lee would have to wear two hats, that of the opposition or Red-force leader, and that of a guide for the battle staff and Aster. Forester's Blue team needed to work fast, to get at the heart of tough yet demanding issues, then deliver options for action that the president could consider. *None of that'll ever happen within the NSC; not if left to their own devices,* he concluded grimly.

The admiral told the Red team's terrestrial cells to consider the ongoing real-world Iranian crisis, widespread Muslim rioting, and the dramatic rise in terrorist incidents. Although many of the recent terrorist attacks seemed to be decentralized and poorly thought out, the onslaught of media images—suicide bombings in public places such as movie theaters, schools, and the like—was striking fear in an already besieged public. So far, no attacks had occurred in the United States but Lee worried about what might be in the works. The cold, fanatical objectives of well-organized groups, including the remnants of al Qaeda, could translate to a deadly strike on American soil.

I loathe this part of Red's work, exposing how truly vulnerable Blue is, Lee thought, allowing a moment of weariness to sweep over him. *But we have to make sure these vermin are part and parcel of Red's strategy. Terrorists are quick, quiet, and well hidden, but very deadly. We can't ignore the wild card of terrorism in this mess.*

Stanton Lee would prove to be a prophetic soul.

14

⊕

THE MALL

About five hundred miles southwest of the STRATCOM wargame center, where a retired admiral was trying to divine the sinister thought processes of terrorist sleeper cells, the fickle Colorado weather had reversed course again. A brisk, blue-sky spring day had turned to winter in a matter of hours, bringing a flurry of dry wind-driven snow to the state's capital. Folks taking an outdoor coffee break on this Tuesday morning suddenly found that they had rushed the season and ducked inside Denver's Park Meadows mall for warmth. Some should have stayed outside and braved the snow.

Near the mall's crowded entrance, no one took notice of a heavily bundled elderly man struggling to roll a battered wheelchair up a handicap ramp, its progress hampered by a thin layer of newly fallen snow blanketing the concrete. A young girl stopped, offered assistance, then graciously pushed the chair up to the shopping mall's oversized doors. Once inside, the man thanked her profusely in heavily accented, broken English. The girl waved and bounded off. She only had a few minutes before her next class began at the nearby junior college, and there was serious shopping to be done.

The heavily bundled old man rolled slowly toward the

center of the mall, eyes sweeping a large river-rock fireplace and lodge-like timbers of the expansive, two-level structure. He rested near a railing, watching shoppers hustle past one floor below. He turned, angling back toward a food court he'd passed. Ah! There, between him and the massive fireplace inside the open court, a group of young children, women, and a few elderly men milled around a cluster of tables. Most were eating fast food bought from the various vendors that ringed the court. Several squealing kids crawled under empty tables, watched half-heartedly by chatting mothers, grandmothers, and assorted nannies.

Hunched over, the frail-looking man slowly rolled his chair toward the group, unhurriedly threading his way among the tables, then stopped near several young women. He smiled, watching the children play tag through the maze of tables. He slowly swung his head left and right. Then he froze. His eyes locked with those of a security guard returning from a restroom near the mall entrance, about twenty-five feet away. The guard nodded. Then his demeanor instantly changed. Something about the old man . . . *Christ! Those aren't old-man eyes!* the guard concluded, alarms sounding within his trained, naturally suspicious brain. He moved toward the food court.

Shariz, no longer the old man, threw the blanket off his lap and leaped from the wheelchair, slamming the abandoned rig against metal tables, scaring three women nearby. Two-handed, he yanked a padded ski jacket open, exposing an odd, wide, belt-like garment jammed with rectangular objects. The security guard was running now, one arm extended at Shariz, the other clawing for his sidearm, and yelling. "Down! Get down!" the guard shouted.

Shariz screamed, *"Allah Akbar! Mag-bar Boyer! Mag-bar America!"* He jerked a wooden handle attached to the belt. A massive explosion erupted, ripping his thin body to shreds and blasting ball-bearings across the court. A shock wave spread outward at the speed of sound, slamming children, mothers, and men against tables, the fireplace, walls and through a nearby store window. Glass shards became thousands of deadly missiles, shredding stacked clothing, athletic gear and the oversized flat-panel screen in the corner of Dick's open-front sporting goods outlet, as well as shoppers and retail clerks, along their flight paths.

For a long few seconds, the mall's food court was silent, consumed by thick smoke and a strong odor of cordite. Then a whimper that grew to a wail erupted from behind a metal table, now on its side, crumpled and jammed against the fireplace's heavy rock base. A small four-year-old girl, blood streaming through dark curls, crawled from a niche between the rock and the table. The stunned child staggered and fell, screaming, "Mommy! Mommy! I can't see! Help me!" But Mommy couldn't help. Life was oozing from the mother's broken body, now crumpled in an unnatural back-arched position across the counter of what had been a Sbarro's pizza and sandwich open-front store. Water from an overhead sprinkler system showered the woman's body, washing blood from her face, eyes open and staring. She didn't move.

Another security guard rounded a corner, running at top speed, then slipped and sprawled across the rock-tiled floor, felled by a layer of glass from what had been a sunglass kiosk and the red slick of rapidly expanding pools of blood. An automatic pistol still in hand, he scrambled to a kneeling position, weapon sweeping the area. He slowly lowered the weapon, then whispered, "Ohhhh, God . . . !" Dozens of bodies littered the area. A few outside the court were moving; most inside were not, tangled among twisted wrought-iron chairs and tables. The guard tried to stand, but his knees buckled. Gulping hard, fighting an overwhelming urge to puke, he finally pulled a radio from his belt and keyed the microphone. His hands were shaking.

"Control, Brantley," he croaked. "We've had a huge explosion in the . . . uh . . . the food court; east entrance, upper level. Multiple casualties! God, dozens! We need medical assistance—*now!* And a hell of a lot of it!" The next few minutes were the guard's worst nightmare come true. Even as ambulances and police special-weapons units converged on the mall, a new danger awaited them. The explosion near the huge fireplace had done unseen damage. The roof began to sag. Its collapse was imminent, but none of the rescuers rushing to the food court horror knew that their lives, too, were now measured in minutes.

Terrorism had struck Denver.

Photos from the mall's wireless webcams surfaced on al Jazeera TV in minutes, prompting fanatical crowds of veiled women, ululating with joy, to spill into city streets from Tehran to Peshawar to Gaza City.

20 APRIL/NORTHERN COMMAND HEADQUARTERS/ PETERSON AFB, COLORADO

Admiral Walter Brohmer ran past the elevator, leaping into the stairway, taking two steps at a time. His aide, an Army lieutenant colonel, was in close trail, wondering what he'd do if the admiral tripped and busted his head wide open. Brohmer landed on the next floor, sprung 90 degrees sideways, and broke into a run again. A major was pulling a heavy door open, clearing a path to the operations center.

Inside, a wall of oversized display screens greeted Brohmer. Phones were ringing incessantly from three rows of adjoining desks and workstations that faced the wall. The Joint Operations Center was filled with dozens of men and women, most wearing olive-drab flight suits, trying to respond to the demanding phones. Several officers and NCOs frantically tapped keyboards, sending data updates to the large-format screens up front. Tension clouded the air, but the noise level was surprisingly muted. These were professionals from each of the U.S. military services, and they were quietly going about exactly what they'd done in dozens of exercises over the last eight years. But this was no exercise.

Brohmer touched a tall, slim Air Force major general's arm, drawing her attention from a computer display. "What do we have, Donna?" He'd been told via phone that an explosion had ripped through the huge Park Meadows mall on the southern edge of Denver, less than seventy miles away from his Northern Command headquarters. Maybe terrorism, maybe not; he'd heard about it only minutes ago.

"Sir, it's starting to look like a suicide bombing. Not confirmed, though. Dozens of fatalities and even more injuries." The two-star NORTHCOM director of operations pointed to a large central screen. "First responders are on scene, and the local sheriff's department is securing the area. A few small fires, but it looks like they're about under control. The mall's being evacuated now, and the cops are clearing the parking lots as fast as they can. No sign of a secondary, but everybody's assuming there's another bomb in the mall."

"Any calls for help?" Even though NORTHCOM's primary purpose was homeland defense, Brohmer *still* had no authority to send military troops in until a federal agency requested them. The fallout after Hurricane Katrina in 2005 had given NORTHCOM more wiggle room to respond in life-threatening emergencies, but he had to wait until the secretary of defense cleared him to act. The restriction on instantly deploying troops, however, was intentional. A Civil War–era law referred to as "Posse Comitatus" precluded U.S. military forces from jumping in and taking charge of a situation like this, a crisis normally handled by local police, fire, and medical personnel. NORTHCOM could provide assistance, and a lot of it, but not until it was requested by the civvies. That point, he recalled wryly, had been made even louder and clearer about five years ago, when his predecessor had clashed with the Homeland Security chief over who'd be in charge of handling the aftermath of a major terrorist attack.

"Yessir. A request's in the works. I've alerted our local Quick Reaction Force, and they'll be ready to roll as soon as you clear them." Brohmer nodded, then reached for his ringing handheld communicator.

"Admiral Brohmer . . . Yes, looks like a suicide bombing, Mr. Secretary. . . . No, but the locals are taking precautions. . . . Absolutely. We'll be on site within the hour. . . . Will do, sir."

The admiral gave his director of operations a thumbs-up and pointed to the big screen, still listening to SecDef Hurlburt. Major General Donna Zurich turned and gave a clipped order to another officer, who nodded and grabbed a phone. NORTH-COM had been activated.

Minutes later, the admiral and Zurich stared at the big screens as one after another filled with data, overhead imagery, TV news coverage, and grainy images captured by webcams scattered around the stricken mall. A news team had a heli-copter over the area, trying to get closeup shots, while avoid-ing police rotorcraft intent on herding nonemergency aircraft clear of the area. Several flight-for-life medical helos were already on the ground, their rotors slowly spinning as bent-over EMTs ran toward the mall's entrance, carrying medical kits and stretchers.

"Holy Mother," Brohmer breathed to himself. Somehow, he felt an overpowering sense of guilt. This was exactly what NORTHCOM had been set up to prevent. All those exer-cises, the hours, days, months, and years spent trying to imagine what a terrorist might do, where a sleeper cell might strike. NORTHCOM's dedicated team had done everything it could think of—within financial reason, he reminded himself—but the sonsabitches had still slipped one through. And they'd hit a crowded shopping mall—one of his personal worst fears—only a few miles from his headquar-ters. *And this could be just the beginning,* he thought, shak-ing his head.

20 APRIL/STRATCOM HEADQUARTERS/WARGAMING CENTER

On the far side of the center, Forester had his hands full, push-ing the primary Blue team to *think,* to develop options that could eventually land on the president's desk in Washington. Aster obviously had strong convictions about the value of in-tegrating Forester's battle staff with the wargamers, but the disruption in normal staff rhythms remained a bit unsettling.

Forester wrestled with himself, wondering whether the effort was worthwhile or ultimately would backfire.

"*What?* An Army grunt 'thinking'?" Aster asked, approaching Forester's group in the wargaming center.

"Some people might say that's an oxymoron, sir." Forester half grinned, stepping to the side. "To be honest, sir, I was thinking about how tough this battle staff integration is turning out to be."

"Oh? What's the problem?" Aster asked warily, an edge to his tone.

"Well, we've been fortunate to pull this group together to help us sort through all these . . . uh, challenges. We get to make decisions in real time, then test 'em before they're actually implemented out there in the real world," Forester explained. "But I'm still uncomfortable with the game's disruption of our normal battle staff routine. Sure, maybe I need to unlearn what I *think* is a staff routine and get back to thinkin' about the bigger picture."

Aster eyed his ops director a few moments before responding. "Yeah, it takes the right people. Normal battle staff functions tend to ignore everything *but* the military side, Dave. Old European experience and ancient Chinese strategies didn't apply to twenty-first-century problems, right? Well, Lee is re-teaching us everything we forgot. And I'm telling you, it applies to what we're up against right now."

Even as he spoke, Aster didn't know the half of it.

"Yessir. We'll owe the admiral a beer when we get this mess settled down," Forester said. "He even insisted that each wargame team leader read the margin notes from the admiral's dog-eared copy of *The Art of War*. Yeah, Sun Tzu makes good sense. But applying it today? At a critical time like this? I guess I'm a bit antsy. Feels like we should be doing something more . . . well, more action-oriented, I guess. Feels like we should be in a global strike mode, dammit."

"Who'd we strike, Dave? Look, Lee's got us thinking way beyond just building good *military* options," Aster said. "That's good for our battle staff people, I think. The civvies in here are forcing us to think through the potential outcomes of a particular decision and action, forcing us to think about what the unintended consequences might be."

"I get it, boss. But still—" lamented Forester.

"Believe me, I understand what you're saying," Aster interrupted. He fixed Forester with a steely look. "But don't be kickin' yourself about what you think you oughta be doing. Not now. I want you to keep pressin' here, in this wargame."

Aster flicked a glance at his wristwatch, indicating the conversation was over. He clasped a firm hand on the slightly shorter man's shoulder, and added, "You and your staff are not wasting your time, believe me. Yeah, we're filing off some rough edges, but it's paying off. I can see it."

Aster stuck both hands in his pants pockets and looked around the room, noting that Lee was heading for the pit. The grin disappeared and for a moment, his tired eyes betrayed the pressure he was under.

"Dave, I need you to get those Blue options developed soonest. We've got a bunch of nervous folks in Washington waiting on 'em. And you might have a quick chat with Androsin. He made a damned good case for merging the battle staff and gamers." Aster winked, then headed for the front row. Forester nodded. *So. Androsin's behind this merger.*

Admiral Lee, now in control of the pit, began his interim Red team assessment, skipping the usual warm-up remarks. Androsin whispered to Jill, "The maestro takes command of the orchestra." Jill smiled and nodded.

"Ladies and gentlemen, the latest Red team assessment is now on your individual workstations, as well as in the registry," Lee began, resorting to his odd, stilted manner of speaking, pausing for emphasis between syllables and, at times, eliding vowels, as if he were delivering a college lecture. Jill had once commented that the admiral sounded like an actor straight out of a 1930s movie.

"I will highlight the key points. And if you will," he paused, "please focus your attention on the center screen." Lee gestured to one of the five large displays ringing the amphitheater.

"Blue's space-system vulnerabilities create a strategic opportunity, and various Red entities are embarked on several fronts to take advantage of what these parties see as a once-in-a-millennium chance to improve Red's strategic position in the world. I'll now enumerate the most likely Red actions that can be expected over the next several days and weeks."

A map of central China appeared on the far right screen, with several sites highlighted by bold flag notations. Satellite images of Chinese military compounds appeared on another screen, correlated with the map's flags.

Lee summarized the outcomes of his Red team's deliberations. "You'll note that the Chinese have four operational sites with ground-based lasers. Two of those GBLs will soon begin test-firing low-powered beams. Blue's intelligence and communication satellites are the targets of these GBLs, we believe. In addition, the Chinese space city at Shanghai will rapidly prep for a manned space launch aimed at improving space-situational awareness. That mission will be followed by three or four launches of intelligence and communications payloads. Those will be from space ports in Hainan and the Gobi Desert. Note that these on-call launch reserves are the manifestations of China's space-strategy initiatives begun in 2002. Chinese leaders also consider all of these launches as purely defensive. I repeat, *de-fens-ive,* actions."

Lee continued to outline the Red team's anti–space-system strategies, switching to those associated with Iran, North Korea, Russia, and nonstate actors. "I know some of you will dismiss the likelihood of so many Red nations and rogue actors pursuing a vigorous counterspace strategy against America, seemingly uncoordinated, but we believe there are plausible reasons for this to occur. One, Red considers an attack on the West's space infrastructure to be low risk, *if* Red thinks he can get away with a *covert* strike. Two, we've already seen the dramatic impact a single rogue maser system has had on the very fiber of Blue's national security. That impact has not been lost on Red, believe me. And three, it follows that continued attacks on Blue's space assets will serve to expose a vital Achilles' heel of the world's so-called *only superpower.*"

Stepping outside his Red-leader role for a moment, Lee added historical context. "I'd like you to recall the two decades of studies, articles, and congressional testimony lamenting how highly vulnerable our space systems are. They repeatedly called attention to our lack of comprehensive space-situational awareness, and our collective failure to develop policies and rules of engagement to deal with attacks on these systems. While the U.S. waffled, any number of Red entities went to school on these subjects, just waiting for the right moment to strike."

Lee swept his right arm horizontally. "That *right* moment is here, now. We must *not* underestimate the importance of this serious wake-up call!"

Aster grinned. *Is Lee lecturing or presenting Red's position? Maybe both.*

A Blue team member interrupted. "Point of clarification, Admiral. Does Red envision any terrestrial action to accompany attacks on the space infrastructure?"

"Good question. You'll find detailed Red terrestrial options in your game registry. And I think we're just finishing up some North Korean items, aren't we?" The admiral received a thumbs-up from a Red player.

"Let me summarize the highlights," Lee continued. "China will increase military preparedness throughout its eastern districts. That includes anti-access deployments against our carrier battle groups; preparation for rocket and air attacks against Taiwan's key military and political leadership nodes, and increased states of readiness among China's strategic rocket forces. Specifically, the latter will be targeting Guam and Hawaii."

The old sailor was stretching reality a bit, Aster felt. But he was still making sense, and it appeared the battle staff troops were starting to understand what the hell Lee was getting at. "On the diplomatic front, China will introduce a UN resolution denouncing space system attacks, recommending punitive measures be taken against any nation interfering with another's space sovereignty," Lee continued. "This resolution is seen as a necessary diplomatic step to protect

China's own space assets, which are keeping tabs on U.S. military activities."

Aster felt a vibration, checked his handheld comm device, and grimaced. He nodded to Lee briefly, then slipped out the amphitheater's main doorway.

As the admiral shifted to terrestrial actions being undertaken by other Red team elements, Androsin and Jill Bock monitored direct feeds from the Red team's inputs. Once BOYD updated the team's moves, the registry's geopolitical and military scene would change to reflect those moves. Other wargame teams could access the new BOYD data and refresh their own situational awareness, then start developing another round of action.

As Jill massaged the feeds from Red at her workstation in the pit, a bloody image almost reached out from the screen and slapped her across the face. She gasped and swore softly. She glanced at Androsin, eyes wide.

"Shit!" she stage-whispered. "Jim! Look at this!" Androsin leaned over to scan her screen, then recoiled. He frowned, read the text-crawler across the bottom, and groaned. Rolling his chair closer to Jill, he monitored the stream of data, images, and text her flying fingers were culling from BOYD's insatiable databases.

"When did *that* happen?" Androsin hissed, aware now that Lee was sending eye-daggers his way. When the admiral was talking, he expected everybody to listen. Jill pointed to a block of text in a new window. Androsin quickly digested the message, then stood. Lee interrupted his dialogue in mid-sentence, pointedly glaring at the Army colonel.

Androsin stepped to the admiral's side, leaned toward the man's white-tufted cranium, and whispered something. Lee stiffened and his face grew dark. He quickly stepped aside, waving the colonel to the pit's lectern. Androsin raised the microphone several inches before speaking.

"Pardon the interruption, but we just got some bad news. Jill, put it on the big screen," he directed. Instantly, the bloody destruction of Park Meadows mall appeared. Emergency crews worked frantically among bodies. The images swayed from one grisly scene to another. A burly green-shirted sheriff's deputy bloomed in the TV camera's lens, both hands raised, firmly ordering a TV-news cameraman away.

"This is live from Denver," Androsin continued. "It appears a large shopping mall was hit by a suicide bomber within the last twenty minutes or so. We'll break now and try to figure out how this is going to impact what we're doing here." He turned and headed for the door, leaving stunned wargamers and battle staffers staring at gruesome images streaming across the large display.

"Those bastards!" exploded Matt Dillon, the Army space-command chief. He turned and ran toward the back door, taking amphitheater steps in giant leaps. Jill's heart sunk. Matt's family was in Colorado Springs. *Oh, God. Please make sure none of them were in Denver today. . . .*

Though stunned herself, Jill forced her attention back to the console, coaxing BOYD's voracious search engines to find . . . what? *Hell, information!* Anything and everything that would help the frozen wargamers understand not only what had occurred in Denver, but why. Then she found it. Actually, BOYD did: streaming images from a half dozen cameras placed among exposed beams in the lodge-like Park Meadows mall. Others peered from inside retail stores, staring at unbelievable carnage beyond shattered windows. One camera lens was cracked, giving its image a bizarre, bisected appearance.

Security cameras! Chain-store webcams for real-time advertising on the Internet, Jill realized. *Damned things are everywhere these days!* The scenes were unfiltered, gruesome, and staggering, in bloodred color and stark detail. *Oh, shit! And they're being automatically uploaded to the Internet!* In an hour, these horror-show views would be flashed to every corner of Earth, seen by millions.

Jill swallowed hard, fighting an involuntary urge to heave. Trying to focus, she weighed the pros and cons, then tapped her keyboard. An instant later, she heard a collective gasp as multiple images, arrayed in a scrapbook-like layout, popped into view on the screens of every workstation in the center. She heard a muffled yelp and glanced up in time to see Preston Abbott, the NSC rep, leap from his seat and waddle toward the amphitheater's rear doors. Holding a hand over his mouth, the man didn't look healthy.

Silence hung over the center as bureaucrats, corporate executives, technologists, and seasoned warfighters—Forester's battle staff—stared in shock at their screens, some glancing up at the big wall display now and then. Everywhere they looked, though, disaster, blood, and death greeted them. Terror had gut-punched every person in the room, civilian and military alike.

Jack Molinero, the TransAmSat vice president, was the first to speak, his voice strained. "How could this happen . . . here?" He looked up, a flash of recognition darkening his pained features, searching the room quickly. "Admiral, could there be a connection . . . uh . . . some link between what we're dealing with . . . ? You know—loss of space-situational awareness, navigation, comm, whatever—and this attack?"

A few feet from the pit, Lee stood ramrod straight, arms folded, glaring at the main big-screen as if it were a personal challenge to him. He nodded slowly, registering what he was seeing, yet mentally logging something in his own "to-do" list. He turned toward Molinero, saying nothing for a long moment. Every eye in the room was now on the retired officer.

"Yes, Jack, I'd say so," he said, quietly, his words laced with firm conviction. Lee flicked a hand toward a corner filled with men and women in uniforms, adding, "Some of our military colleagues may recall a key element of the first DEADSATS wargame. It was underplayed in the final report, in my estimation, but that element distinctly illuminated how the loss of friendly space assets would be seen as a golden opportunity by some of the world's unsavory characters. Tenuous it may seem, but there are undeniable links among the current degradation of America's space infrastructure, our response to those losses—such as ill-advised action against Iran—and what appears to be a terrorist bombing in Denver.

"No question about it," he concluded. "An astute terrorist network now sees an opportunity to unleash havoc and pain within the heart of this nation. I fear this is only the first strike, ladies and gentlemen. We can expect more of these horrors, and soon."

15

COMING TO GRIPS

The American people were scared, incensed, and frustrated. Scared by a suicide bomber attack deep within the nation's heartland, and scared of what that might mean for their families' safety. *Denver, for God's sake! If Denver, then it could happen anywhere. In the mall where I and my children shop. The office building where I work. My kids' school!*

In the days following the attack, business slumped markedly as people holed up, leaving their homes only for the necessities of day-to-day existence. Absences at school, work, and civic functions skyrocketed. Fear ruled from east to west and north to south. American city streets became eerily silent after dark, devoid of casual traffic, and as desolate as Main Street in a deserted mining town.

A collective paranoia gripped America, a fear of anything strange or out of the ordinary. People watched one another too closely because everyone was on edge. Mistrust was everywhere, making neighbors wary of each other and parents even more mindful of where their children spent afternoons after school. The presence of slow-moving police cruisers along suburban streets and lurking behind the oversized Dumpsters at shopping malls also made people feel as if some all-knowing, all-seeing entity was scanning for the slightest indication that someone was not walking a straight line of normalcy. It was unnerving.

Anger, too, ran deep, crossing political, gender, religious, and economic lines throughout the country. Not since the terrorist strikes of 9/11 had Americans been so united in their outrage against attackers from the outside. The mall bombing, coming on the heels of their president's grim prime-time announcement that all astronauts aboard the International Space Station had died from a maser attack, shocked Americans to their core.

Shock also turned to seething anger when, shortly thereafter, the link between Domingo's drug cartel and a shadowy, high-ranking figure in Iran's theocracy had come to light, despite the White House's best efforts to keep that link deep-black secret. But now this attack, deliberately targeting children in a shopping mall food court, turned anger into rabid fury. This time, the bad guys would pay more dearly than anyone had ever paid for an attack on America.

But there *were* disagreements over who was responsible for the attacks, and what America's responses should be, particularly between seasoned intelligence and military personnel and the White House's political appointees. Paul Vandergrift, the national security director, accused the Pentagon of leaking information about ties between the cartel's maser team and Iran's leaders. In turn, SecDef T. J. Hurlburt privately aired his own criticisms—and suspicions—about "damned NSC cowboys" launching *Sting Ray,* then sending Charlotte Adkins into the lion's den to get executed through a naive, senseless, off-the-books operation that undermined U.S. interests. At least the NSC's incredible double-shot stupidity hadn't hit the press . . . yet.

Not that it would have mattered much; things couldn't get much worse for the president, it seemed. A swelling outcry of rage from Main Street, Everytown, was resounding through the halls of Capitol Hill, echoing down Pennsylvania Avenue, and rolling over the Boyer administration. Mutterings about impeachment were already being made on the Hill and by conservative talk-radio hosts who were whipping up already angry drivers during morning and evening commutes, reinforced by nightly cable-TV commentaries.

The perception that paralysis and impotence gripped official Washington fueled citizens' anger and led to open demonstrations of national frustration. Even the decidedly liberal East Coast media television and newspaper giants, the very entities that had helped sweep the dark-horse Virginian, Pierce Rutledge Boyer, to victory in 2008's presidential election, were bending to the overwhelming pressure of public demands for retaliation. Their editorials, initially tentative, questioned the administration's apparent dithering, but were softened by calls for "international pressure" and the proven-toothless strategy of "sanctions against Iran."

Finally, the usually loyal California senator C. I. Creighton, Democratic chairman of the Senate Armed Services Committee, publicly questioned the Boyer administration's lack of response to the ISS attack. The Santa Monica firebrand clearly smelled Boyer's blood in Washington's waters, and was putting distance between himself and the White House.

You SOB! Boyer flared silently, staring out the Oval Office's expansive window early one morning. He held a copy of the *Washington Post* editorial page in one hand and a cup of coffee in the other. Creighton's op-ed piece in that paper called on Boyer to "get out from behind his desk and lead! America is demanding a response to these egregious attacks." *Always jockeying for self-advantage, especially when the going gets tough. Loyalties be damned. Creighton, I won't forget this, you bastard!* Boyer resolved, tight-lipped. In disgust, he tossed the newspaper into a wastebasket, then turned back to the window. Not really seeing the groomed White House grounds, he sipped the Italian espresso roast that kick-started his every day.

But the papers were right, Boyer admitted. He needed a homerun, and soon; something to defuse increasing global tensions while simultaneously reassuring voters that he and his administration had a firm hand on the national-security tiller. *NSC's a mess!* he snorted. He couldn't expect innovative, aggressive solutions from most of those eunuchs, who had already made a bad situation worse. He shook his head, still finding it impossible to believe Paul Vandergrift's team had screwed up so badly in Iran, fanning the flames of Islamic fervor.

Those stars and bars out in Omaha . . . By God, they'd better come up with some profound options for action! And soon. If they didn't, America's position in the world and citizens' confidence in the country's national security forces, not to mention Boyer's own political career, were doomed.

21 APRIL/STRATCOM HEADQUARTERS/WARGAMING LABORATORY

"You know, Jim, I don't think Blue is keeping up with what's happening in the real world. I mean, look at this stuff coming out of Red! It's like they're reading tomorrow's news today! They're running rings around Blue!" Jill said, waving a stack of BOYD printouts. "Al Qaeda attacks against our satellite ground stations in three different countries. Suicide bombers in an inflatable raft taking out an oil rig in the Gulf off the Louisiana coast. Two attempted attacks against petro-chemical plants in California by 'persons unknown.'" She could have continued, but Androsin's glower as he focused on his own workstation screen stopped her.

"I know, Jill," he said finally, moving his finger across the bottom of a touch display. "It's a hell of a list. But we're doing . . ."

"Hell, it's not just a *long* list!" she sputtered, barely able to contain her own frustration at Blue's seeming inability to grasp the magnitude of threats. "There're some damn-near fatal blows in here, assuming Red's projections all come true. I mean, just look at this from Lee's Red team: *'Iranian and North Korean nuclear strikes coupled with cyber-attacks . . .'* And what about these suicide bombers? One's bad enough! What if they really *do* start hitting us all over the country? It's horrible, just horrible! And Washington's taken its eye off the ball. Those sonsabitches are doing nothing! What the *hell's* going on back there?"

Androsin glanced up, taken aback by Jill's salty outburst. "Hey, don't get your undies in a twist. You're just focusing on what the public sees, like suicide bombings and gory scenes splattered all over the tube and front page. The friggin' media's fanning the flames with all that blood and guts. Damned media!"

Jill hesitated, then cocked her head and eyeballed Androsin. "Hey, Colonel! *That's* what we're missin' here."

He paused and noted Jill's wide-eyed look, as if she were seeing through him. When her voice changed just so, and that . . . that thousand-yard stare appeared, he knew she was on to something profound.

"Missin' what? Tell me something I haven't seen," Androsin challenged wryly.

"The media!" Jill said, her eyes aglow, as if on fire. "Don't you see, boss? We're takin' a hammerin' because no one's tellin' *our* side of the story. Al Jazeera's outgunning us up and down the field. We're missin' an opportunity to have the media spin things in other directions. We need to get some of *our* stuff out there, into the light of day. Provide some transparency as to what we're doing here. The media can help us do that."

"Jill, that's nuts! You know how classified this game is! Besides, we've got public affairs feeding the newsies enough as it is."

"Okay, okay. I know most of what we're doing is classified, but don't tell me our public affairs officers really know the media. They're nice enough folks, I guess, but they really don't have much clout with the major wire services and such. That's why they're called flacks. And the stuff our PAOs put out is gross pablum. The lazy reporters pay it lip service, because we give them at least limited access. But the good ones are always sniffing around, trying to find someone on the inside who'll give 'em a whiff of what's really going on," Jill said, waving her hands and becoming more animated.

"But then they're like the blind men and the elephant. They grab a tail and conclude *that's* what the whole elephant looks like, long and stringy on one end and a big hose on the other. What do you think's creaming the Washington crowd right now? Bits and pieces leaking out! But nobody in the media has a good, full-view perspective on what's really going on geopolitically. So, the Beltway wonks are running around like chickens, trying to put out little news brushfires caused by incomplete info or outright misinformation spewing from the White House and the Hill. And the poli-wonks are losing the perception battle, always playing catch-up!"

She left her station, pacing now, staring past Androsin, thoughts cascading ahead of speech. "Think about this! We bring in some heavies from the media, kind of a super press pool. They get to hear our teams' briefings following each wargame deliberation. No BOYD access, of course; just the briefings. We could treat 'em like war correspondents— embeds; let 'em follow the wargame as it unfolds, seeing what goes into each move and getting a feel for the range of issues being addressed by our civilian and military gamers. Then our public affairs guys do what they *are* paid to do: make sure nothing gets released that could endanger national security or endanger lives."

"I still don't see any advantage to havin' the damned media here," Androsin groused, shaking his head in doubt.

"Boss, stay with me here. We help selected news outfits focus on the bigger picture, so they understand what it all means in the larger sense," Jill explained. "Suicide bombers are the headliners now because *that's* the story, from the media's perspective. But those bombers are like rioters. See some chaos, and decide, 'Hey! Let's join the fray!' You've seen the intel. These bombings are small, one-off hits. Nothing really organized, and the protagonists aren't going after strategic targets. With the massive crackdown by law enforcement—aided by NORTHCOM's military support, of course—they'll fizzle. There just ain't a supporting infrastructure for these terror freaks here in the U.S."

She paused, re-forming her arguments. She knew she was on to something, but hadn't sold it yet. They were into a *War of the Worlds*–type scenario all over again. Tell the people they're panicking and they panic. Stress the ineffectiveness of the enemy, instead, and it'd build confidence. The media naturally lead with whatever bleeds. Everyone knows that, especially the bad guys. So the trick is to make the *enemy* bleed, in print, without telegraphing your punches.

"You gotta know how the media works, boss," she finally resumed. "Then work the media."

"I got that part," Androsin nodded, starting to understand, finally getting into sync. "The story that's not being told now is how the United States is purposefully going about searching for peace-building options, while ensuring that we're ready for war, of course. Americans need to see that we're working hard on peaceful solutions, not just looking for war options that'll *command* the peace."

"Right-on, boss!" Jill said excitedly, jamming a finger toward the colonel. "And we know bad boys like Iran and North Korea won't care, but China will. They do business here, they're major-league investors in our economy and have been heavy-duty manufacturers in Asia for the past decade. 'Money talks,' and they know it. If we can show China more of our real intentions, through the media, and they buy it, we dodge the big, granddaddy go-to-war-with-China bullet. That'll give us more time to deal with the crazies that are *totally* unpredictable."

Androsin started to respond, but she cut him off. "One more point, boss. There's precedent for what I'm suggesting. At the Harvard School for Government, we studied something called 'the presidency and the media.' Back in the fifties, Eisenhower's staff and the Pentagon gave in-depth background briefings to so-called trusted reporters. Even classified stuff! Apparently, the strategy was quite effective. When something really big and sensitive was going down geopolitically, at least that handful of reporters had the big-picture perspective, could read between the official lines of B.S. and understand what was really going on and the stakes involved. Their stories were more measured, making sure the right messages were sent to friend and foe alike. At times, those news articles were more effective at defusing really tense situations than high-powered, diplomatic wrangling behind the scenes. I think this might be just the right time to break our present-day media mold and re-create an old one: the Eisenhower strategy."

Androsin nodded, a grin starting to spread. "Sold, Jill. When you get a minute or two, like within the next hour, scratch out a one-page point paper about this. I'll get it to the Big Boss ASAP. I think you've come up with something that could make a huge difference here."

Androsin's communicator buzzed, sending Jill back to her computer. A few clipped sentences and the colonel headed for the door, calling over his shoulder, "General Forester's hit a snag. I'll be in the amphitheater." Jill waved and continued working.

Back in the wargaming center, Androsin read Forester's frustration immediately. "Sir, you wanted to see me?"

The general glanced up from studying Blue team updates. "Uh, stand by, Colonel." Forester posed an open-ended directive to the group, his tone cutting. "Look, people. We gotta get out ahead of the problem here, but let's stay on track, too. Remember, our primary objective is to avoid anything that could widen the war we already have on our hands. Things like more space-asset losses and all sorts of terrorist attacks. Focus on those issues first, *then* we'll work on deterring or destroying Iran's nuke capability."

Androsin nodded, but his frown reflected concern. *General's a smart guy. But he's still thinkin' like a staff puke. This whole wargaming process is s'posed to improve in-parallel planning, forcing the staff to consider multiple issues in preparing for a range of actions.* The hard-core gamers were helping Forester and the battle staff see that. But Forester's priorities were out of whack. *Blue had better deal with those Iranian nukes, and pronto. That's one sure-fire way to avoid a wider war!*

Forester turned to Androsin, eyes heavy with fatigue, jaws tight. "Colonel, you've been doin' this gaming stuff for some time, and I know you're comfortable with it. But we're strugglin'. My battle staff is still all over the place, and getting nowhere fast, as far as developing options for the president. We may have to rethink this whole approach."

Androsin nodded, but left a heavy pause hanging for a long moment. Forester's support had been tenuous from the outset, and even that was clearly waning. If the operations director and his staff bailed out of the wargame now, both Forester and he, Androsin, would have to answer to Aster. Worse, they'd lose a golden opportunity to improve and accelerate STRATCOM and national security community decision making.

Integration of the exceptional knowledge bases inherent in these wargamers and the general's battle staff was already paying dividends. But Forester didn't see it. Androsin knew they were making great progress, seeing positive changes already, after only a few days of integrated, intense work. He had to convince Forester of that, though.

Forester was in a foul mood. "And exactly whaddya smilin' about, Colonel?"

"Oh, sorry, sir! Just thinking about how your staff is working with the gamers. In a way, the battle staff is acting like independent assessors, like the old umpires in our not so distant wargaming years."

"Colonel, I can't afford to change their roles so damn much. We've got serious strategic issues to address, and I'm not at all convinced my staff's spending its time productively here." Forester was a whisker away from yanking his staff out of the game.

Androsin drew a deep breath before responding. "Sir, I can see how you might come to that conclusion. But you and the staff are making a hell of an impact, and you're making solid contributions to what really needs to be done right now. In fact, you're *in* the war right now. Your people are embedded with the wargame teams to such a degree that they're beginning to think differently, gaining the broad understanding needed to work the gamut of military options in a much wider context. They're gaining the perspectives needed to avoid those perennial unintended consequences that traditionally bite us Americans in the butt. Your staffers are already fighting the war and winning. They just don't know it yet."

Androsin paused to take a reading of the general's

demeanor, trying to sense whether he still had the man's intellectual attention. He did. Absurd as it might seem at the moment, Forester was winning, albeit without realizing it, simply because his staff was planning differently.

"The real power of this approach goes beyond that," Androsin continued, diving deeper. "You've already got a much smarter battle staff in terms of cognitive reasoning. I'd suggest letting this first major move unfold and see how your people do. Sure, Red's all over you, but look who you're up against, Admiral Lee. He's making Blue fight harder, smarter. Your folks'll be ready to brief their findings soon, and I think they'll surprise you with the depth and quality of recommendations. Based on what I'm seeing, they'll water some eyes in Washington, too."

Forester was nodding, but his crossed arms still screamed skepticism. "Yeah, the whole interagency dialogue is significant, plus we have these industry guys and the academic types. But, hell, the NSC should be running wargames like this! An integrated environment like you've built here, but at a higher level. We're doin' what *those* SOBs should be doin', and we should be acting on what *they* come up with!"

Androsin nodded. *Damned straight. Too bad the NSC has been hiding inside its own silo for the past forty years. Idiots!*

"If I may, sir . . . ," Androsin offered. Forester shot him a grim glance of impatience, but said nothing. "What if your team backs up and takes another look, but from a 40,000-foot perspective. You're right. There's definitely a possibility we'll lose more space assets and take more terrorist hits, but those issues are being addressed by others. The president can't really do more than what's already being done."

"So, what the hell are you trying to say, Colonel?" Forester barked.

"Sir, maybe your Blue team—and your battle staff—should be taking a hard look at what Red's generated with respect to the most threatening big-picture situation the president's now facing: Iran," Androsin said, knowing full well he was tweaking the general's ego. "What measures could the U.S. take to ensure those fools don't launch another nuke? How do we whack them for executing a Canadian emissary sent in peace, even if it was actually a rogue NSC op? How can Iran's nukes be neutralized in such a way that the president doesn't trigger an all-out war with Iran—and probably the whole blasted Muslim world, too?" The colonel shut up, deciding he'd said enough.

Forester looked hard at Androsin, then nodded absently. *He's thinking again. Good* . . . the wargaming chief concluded, relieved.

The STRATCOM Deputy J3 for Global Strike, Nate, suddenly appeared and interrupted. "Pardon, General. I've completed an assessment of Red's next moves. You wanted it presented to the Blue team for review and development of courses of action as soon as I had it ready."

"Roger. Let's get on with it." Forester keyed his microphone, addressing everybody on the Blue team channel. "Okay, listen up. Nate's got the Red move inputs. He'll walk us through 'em, then we'll discuss 'em, drawing on BOYD's analyses. And let's start thinking strategically, people. In particular, I want to hear some ideas about how we handle Iran." Forester nodded toward Androsin, a half smile saying: "Message received."

The colonel stayed with Forester's Blue team a while, listening to its characterization of the most immediate threats to the United States and its allies. The team's dozen or so members debated the utility of diplomatic initiatives, then built a range of Iran-related response options for the NSC and president.

Nate, the deputy J3, recapped the satellite losses and their consequences. He then painted a grim picture of potential Chinese and North Korean actions, exacerbated by additional terrorist attacks at home and abroad. He outlined the escalating Iranian crisis, emphasizing that the combination of space-system losses and the NSC-initiated IO-worm attack against Iran, exacerbated by the ill-fated Adkins mission, was having far-flung impacts. Some potential outcomes were unpredictable, involving an array of possible state and nonstate players. Clearly, the U.S. space infrastructure degradation and the NSC's missteps were enticing other potential adversaries to join the fray.

Forester then had a Blue team member enumerate response options. The usual courses of U.S. action came to the fore: Deploy two carrier battle groups to Northeast and Southeast Asian regions, one to defend Taiwan, and a second to protect sea lanes through the Spratly Island choke points. Station ballistic missile defense ships in predefined Japanese, Guam, Alaskan, and Hawaiian "intercept baskets." These were the areas having the highest probabilities for intercepting and destroying incoming enemy missiles. Convince China to remain on the sidelines, and enlist Chinese pressure to rein in North Korea. Initiate a *démarche* against North Korea, warning its dictatorship of grave consequences should it begin large-scale troop movements out of garrison, particularly if accompanied by an increase in strategic rocket forces' readiness. Conduct B-2 and F-22 air strikes against Iran's C2 network and nuclear-weapons development and storage facilities.

As the briefing of key Blue options continued, Androsin thought he sensed Admiral Lee's guiding hand. With Nate, the deputy J3, acting as an observer during Red team deliberations, Lee had created an effective information shunt to the Blue team. *Makes sense,* Androsin thought, admiringly. The magnitude of potentially negative impacts due to Red actions during Blue's most vulnerable time warranted such an info back-channel.

A cryptic Blue option-item caught the colonel's attention, recalling something Jill had said: "Sooo, we've got a spaceplane to play with." Androsin reread the item on the big screen: *Conduct covert strike against Iran's nuclear capability, specifically the underground nuclear facility at Parchin, Iran.*

"It's working," Androsin said to himself. *By God, I think they've got it! The president* will *get some good options. . . .*

21 APRIL/STRATCOM HEADQUARTERS/COMMANDER'S OFFICE

Hunched over the STRATCOM commander's conference table, Forester finished summarizing Blue-developed options, then looked from Aster to Lee. "We've got a lot to work on here. Any last-minute thoughts, Admiral?" Aster was reasonably sure what the old sailor would say. In fact, the STRATCOM chief had already set the preflight wheels in motion.

Lee stood and crossed his arms, pacing slowly around the office, head down. Aster thought the old Navy officer was showing signs of fatigue, too, his normally square shoulders slumping. The troops were wearing down.

"Yes, I do," Lee began, speaking carefully. "Iran's nuclear capability must be destroyed without hesitation. But it's imperative that the strike be totally covert. Your spaceplane is a viable option that needs to be carefully explored. *Anything* we do against Iran demands absolute, airtight deniability, though. Global sensitivities and anxieties are simply too extreme to risk having American fingerprints on such a mission." The two generals exchanged a glance. Lee caught it, then cracked a rare smile. *Good. As I suspected. The mission is probably already in progress.*

16

RODS FROM GOD

The secretary of defense had dropped the issue on the table where the president could no longer ignore it. No pussyfooting, no political niceties, no milspeak. Just a straightforward military option to take out Iran's nuclear threat. He waited for the president to decide: yay or nay? With Congress and the media clamoring for his head, what would he do?

"It's risky, but I think it's our best course of action, sir," T. J. Hurlburt concluded. He watched the president carefully. *Show us what you're made of, Pierce. Warrior or wuss?*

Paul Vandergrift started to say something, but a steely glare from Boyer made the NSC chief snap his mouth shut. He was on very thin ice, thanks to the Charlotte Adkins and *Sting Ray* fiascos. Politically, he might not survive those massive stumbles.

Hurlburt struggled to squelch a smirk, but couldn't resist regarding Vandergrift with open disgust. He hoped to heaven that the NSC weasel would be cleaning out his desk before the week ended. *The damage you've done to this nation . . . You oughta be tried for treason, bastard!* The SecDef seethed, in silence.

Boyer asked a few perfunctory questions, some of which were purely technical. He was covering his bases as best he could. He needed a hit, to land a punch, but without looking desperate before the rest of the world and the media. So, he asked more questions, hemmed, hawed, struggled with a few details, accepted clarifications from the small group surrounding the elegant Oval Office coffee table, then stood abruptly. The others followed, knowing the session was over.

"T.J., you're cleared to launch the mission. It goes without saying that there *cannot* be a single American fingerprint on this operation. Make sure your people fully understand how critical that point is. And best of luck to Speed Griffin. There's a lot riding on his shoulders," Boyer said. He turned to his desk, leaving the group to quietly depart.

As they passed through an outer foyer, Hurlburt leaned close to Vandergrift, grasped the man's right bicep, and muttered, "Paul, don't even think about sticking your nose into this one. The nation can't afford another of your royal screwups. If I smell your stink anywhere near this mission, I'll personally cut your gonads off and stuff 'em in your mouth. *Do we understand each other?*" Hurlburt hissed, squeezing the NSC chief's arm harder than necessary, eliciting a flicker of pain across Vandergrift's eyes. The two stared at each other for a long moment, then Hurlburt slowly released the arm and stalked away.

Vandergrift shuddered slightly and glanced about. Boyer's assistant was busily studying a document, and the president's chief of staff was walking well ahead of Hurlburt. The NSC director turned toward his office, emotions racing from biting, purple anger to a gut-gripping fear he couldn't shake. He'd *never* been spoken to in such crude terms . . . but the SecDef's threat had unsettled him. He would let this one go. Climb on board if it succeeded; plunge a knife between Hurlburt's shoulder blades if it did not. He needed to compose himself.

He needed a drink.

22 APRIL/160 MILES ABOVE THE PACIFIC

Speed Griffin tipped his head back, relishing the view and marveling at his good fortune. The god of flight had smiled on him yet again. He was back in space, flying the XOV-2 for the second time in a week. Launch from the SR-3 mothership and climb to orbit had been flawless, an amazing feat for an extremely complex system that had been in flyable storage for the best part of a decade.

"Gaspipe, Control here. We show you . . . uh . . . seven minutes from the basket. All systems are a go on this end." The ground control team's cramped quarters at Groom Lake were overflowing today, Griffin reflected. *One hell of an audience. Another damned good "opportunity to excel,"* the one-star pilot mused perversely. That little ditty had become a standard mantra among pilots and military leaders who found themselves in very difficult life-or-death situations, a dark, backhanded reminder: *You can't afford to screw this up, bud.*

"Rog. Gaspipe copies. Stand by for prelaunch setup." Griffin's hands flew across the XOV-2's instrument panel and left-side console, pushing switches and cross-checking alphanumeric readouts on several screens. A glance in his canopy-mounted mirror confirmed what the graphic displays had already told him. The trapeze-like rig had deployed from the XOV's Q-bay, extending well above the winglike bay doors that had hinged open and extended almost parallel to the blended wings on each side of the fuselage. Two long, dark, deadly-looking cylinders were locked on sturdy rails that ran longitudinally, pointed in the same direction as the spaceplane's nose.

"Two minutes to basket, Speed. Prelaunch conditions all green," the controller's voice said, tension adding an edge to every transmission.

"Gaspipe, confirmed. Systems in position." Although communications to and from the spaceplane were encrypted, the extreme sensitivity of this White House–approved mission dictated such nondescript terminology. Griffin now focused every fiber of his being on nudging the XOV's nose to a precise attitude. Small puffs of gas fired by vernier rockets and exhausted through machined ports ringing the vehicle's nose, wingtips and tail section answered the pilot's commands, relayed via right-hand sidestick controller. Finally, a crosshair-like symbol on a central display directly in front of Griffin aligned with a bright red rectangular box. The triple-redundant fire control computers had their targeting solution.

Griffin swallowed the butterflies that threatened to jam in his throat, and lifted one shoulder, then the other in an attempt to relax tight muscles. *Breathe, dammit! Just breathe . . . like shootin' a free-throw on the basketball court. Breathe, relax . . .* It didn't work, though. What he was about to unleash was a first for humankind. And the stakes were humongous.

The green countdown digits on his targeting display were into single numerals. "Stand by for deployment," the pilot clipped. No response from the ground, per the plan. Until this was completed, it was critical that the command downlink stay open, just in case the mission turned to crap in a heartbeat and he only had time for one quick "Alpha-Sierra" transmission. A quick scan of several readouts showed "the package" was hot.

Three . . . two . . . one . . .

"Deploy!" Griffin croaked, silently cursing the catch in his throat. A muffled roar and flash of brilliant light streaked past his canopy as a highly specialized missile left the Q-bay rails and quickly disappeared, lost in Earth's palette of bright colors. Still upside down in relation to the planet, Griffin double-checked his display, ensuring that a new icon representing the "Rod-from-God" missile streaking away from his spaceplane was within the target box's red borders. It was. A new set of red numerals was counting down now, still in triple digits.

"Fox one," he whispered to himself, unable to resist a fighter pilot's traditional "missile-away" call. Griffin glanced at a second set of green numbers, verifying they, too, were quickly edging toward zero. His right index finger tightened again, then squeezed as the green digits read "zero." Another

bright flash and a discernible "whoosh" rocketed past his helmet, its white-hot light still visible behind his briefly closed eyelids. He caught a glimpse of the missile's fiery tail, then it, too, was lost in Earth's myriad shades. A glance confirmed two dancing-diamond icons were well inside the red target box. *Not much we can do now. Either Jekyll and Hyde do their job, or they don't.*

"Control, Gaspipe. Packages deployed and on the yellow brick road." Only double-clicks sounded in his headset, acknowledging his radio transmission. Obviously, the troops down below were focused on telemetry signals beamed from the two missiles, relayed through his XOV to ground stations. Griffin exhaled, switched a tiny forward-aimed digital video camera-recorder system off, then set to retracting the trapeze-launcher rails.

Within seconds, a reassuring green light glowed, indicating the Q-bay doors were firmly closed. If they hadn't, safe re-entry would have been impossible.

"Control, how're we looking for Phase Two?" Griffin radioed. They might be locked onto the missiles down there, but he still had work to do up here. And if he missed the next window, the second objective of his mission would be lost.

"Lookin' good, Speed. You're cleared for Phase Two setup. And your first deliveries are bang-on." Griffin winced. That was a little obvious, not exactly approved comm protocol for *this* mission! He quickly reconfigured the XOV-2, selected the equivalent of a second set of waypoints on the autonav system, engaged it and sat back. *Over to you, George. Take us where you will.* He craned his neck, trying to see the expanse of sand and mottled green that he had just targeted. He had no illusions that he'd see anything untoward, nothing to indicate the missiles had succeeded. *Unless there's a secondary. God help us if there is . . .*

The two missiles plowed into the ever-thickening atmosphere, their GPS-aided inertial-guidance systems registering Mach 25 initially. Dubbed *Jekyll* and *Hyde* by the gallows-humor minds that had designed, then hand-fabricated the unique projectiles, they were essentially heavy titanium rods fitted with fast-burning solid-rocket motors, a guidance package, sophisticated sensors, and an ablative shield. Technically, those highly modified shields were part of USAF Space Command's Common Aero Vehicle or CAV family, but the missiles they protected would never be found on any list of CAV payloads. In fact, these were only the second and third "Rods-from-God" ever fired from space. Griffin's XOV test-pilot cohort had logged the first test-shot from space years ago. It had been a raving success.

Only seconds apart, the missiles slowed very little in their long, arcing trip to Earth. Their shields burned away on schedule, enabling tiny pulses of gas to keep the dense, compact projectiles "in the basket," an imaginary cone that funneled down to a point in the desert sand.

Below 40,000-foot altitude, still at high hypersonic speeds, the missiles' terminal-guidance systems activated. Telemetry still flowed as a string of data bits back through the XOV high overhead, giving engineers half a world away a running account of the weapons' health and trajectory.

The dusty sand screamed upward, filling the engineers' screens with grainy images warped by shockwaves and aerodynamic heating that distorted the tiny fisheye optics in each missile's nose. "Come on, Jek. Hang in there!" an engineer whispered to himself, breaking a thick silence that gripped the Groom Lake control room.

In an instant, the first projectile glimpsed a berm of sand dotted with small-diameter, pipe-like stacks protruding. Then the image disappeared. Jekyll had impacted. Eyes flicked in unison to the other missile's screen. Its final images, transmitted only seconds before the second missile's impact, displayed a plume of sand erupting from the sizable berm, followed quickly by a massive explosion. Hyde dived straight into the throat of that hell, its image blinking out, as well. Silence continued to grip the control room, its occupants still staring at electronic fuzz and static that now dominated both screens.

From a corner, a senior officer held a phone receiver tight against his right ear, then raised one hand, drawing the attention of several others. "You're sure! Have you cross-checked that?" He nodded, listening, then smiled broadly. Still holding the phone, he turned and shouted, "Space Command's confirming a large IR signature in the target zone." He paused for the second confirmation. "With a double peak!" he shouted. The room's silence erupted in a collective roar. Fists thrust skyward were quickly converted to high-fives.

"Gaspipe, Control. Be advised: The Eagles just scored, and they nailed the point-after conversion, as well."

From miles above, Griffin nodded vigorously. "All right! Go, Eagles!" He could hear considerable commotion in the background, which could mean only one thing. Both the Defense Support Program and the newer Space-Based Infrared System satellites more than 22,000 miles above Earth had detected a sizable heat signature in the vicinity of Parchin in the outback of Iran. Chances were good, then, that the Jekyll and Hyde "Rods-from-God" had hit Iran's primary nuclear-weapons development and storage site.

"Control, Gaspipe here. Anything from our friends in Dolphin-land?" The Air Force Technical Applications Center's headquarters was at Patrick Air Force Base, Florida, not far from Miami, home of the Miami Dolphins pro football team. The secretive AFTAC organization monitored "nudet" or nuclear-detection sensors that rode piggyback on DSP, SBIRS, and even GPS satellites. If the IR flash at Parchin had triggered a nuclear blast, they'd see it. On the other hand, if the Iranians had done their homework properly, any nuke weapons stored there should *not* have detonated.

"Gaspipe, that's a negative. Dolphins are still zero-zero . . . but we're still in the first quarter. Maybe later," the controller responded cryptically, maintaining the football code they'd agreed to use throughout the Iran-strike mission.

Griffin had been briefed that AFTAC would deploy one RQ-4B Global Hawk unmanned aerial vehicle at 60,000 feet downwind from Parchin, yet outside the Iranian border.

A couple of stealthy, highly classified "Stalker" drones provided by the National Reconnaissance Office cruised at 10,000 feet, tracing the same oval-shaped pattern as the Global Hawk, but at a much lower altitude and *inside* Iran's border. If any nuclear debris was carried from the site by forecast winds, sniffer systems on those platforms should detect it, enabling samples to be captured for subsequent analysis.

Over its six-decade existence, AFTAC's airborne sampling fleet of manned aircraft had snagged nuke debris from Soviet, Chinese, French, and other nations' nuclear tests. Those samples had provided hard evidence about the technology of those weapons—valuable intelligence that helped keep the Cold War cold. Today, though, AFTAC's unmanned aerial vehicles were simply looking for potentially dangerous leakage of radioactive materials blown into the atmosphere. The unit already had samples of one Iranian nuclear weapon, thanks to that recent near-miss Shahab-4 attack on the U.S. base at Aviano. Identification would be easy this time.

"Gaspipe, Control shows you approaching your Phase two orbit. Should be getting a cross-range burn in . . . thirty seconds."

"Rog, Gaspipe concurs," Griffin said. He monitored an autonav-driven graphic on his primary flat-panel display, noting that a stub-winged icon representing the XOV-2 was, indeed, merging with the new-orbit line. His hands moved to the throttle and sidestick controller, ready to take over if the autonav-commanded burn went awry. He felt the brief rumble of powerful aerospike engines and noted the spaceplane's attitude change as vernier rockets also fired. He waited, eyes scanning instruments and displays, until the powerplant and attitude-control systems settled down.

"Control, Gaspipe shows on-track. Confirm?"

"Gaspipe, we show you nominally on the Phase two orbit; objective is 13.6 minutes out."

Griffin double-clicked the mike switch on his throttle, then started configuring the forward-looking video camcorder system again, this time in full-zoom mode. He also selected one instrument-panel screen to display the camera's video image in real time. If this maneuver worked as planned, the camera should pick up his targets long before he could see them with his Mark-1 fighter-pilot eyeballs.

He twisted in the cramped, form-fitting seat, struggling to reach a compartment behind his right elbow. Even though his pressure suit wasn't inflated, its bulk still made moving difficult. And he couldn't see well over the lower lip of his fishbowl helmet. Fumbling awkwardly by feel, he finally got the compartment open, freeing a Nikon D2xi digital camera. Still out of his line of sight, the camera bypassed his hand and floated upward, ricocheting off the canopy's transparency before he could grab it. *Damn! Scratched the flippin' canopy!* In an airplane cockpit, a dinged canopy was no big deal. Spaceplane canopies were a bit more sensitive, though. That small camera-dug divot could become a weak spot during the high dynamic-pressure stress of ascent or reentry. Nothing he could do about it now, though.

Griffin activated the Nikon and prepared it for handheld shooting, while keeping an eye on the time-to-go digits counting down on his autonav display. *What I'd give for a decent radar!* he muttered. The spade-shaped nose of the XOV-2 wasn't exactly conducive to installing a flat-plate radar antenna, no matter how compact, nor a composite radome transparent to radio-frequency energy. So, without a radar, locating a target in space meant going back to basics: drive over there, and visually spot the things.

"Gaspipe, Control shows you within one minute of rendezvous. They should be approaching overhead, Speed," the test engineer said. Now right side up in relation to the Earth, Griffin searched the dark of space above his canopy. His orbit was lower than that of his objective, so he was going faster, too. He'd literally fly under the target, then add power, which would raise the spaceplane's orbit to match that of his objective. If he did it properly, he would literally back into position with his target.

Then he spotted them, the Chinese nanosats that a USAF space surveillance squadron inside Cheyenne Mountain had tracked during their launch a few weeks ago. However, the tiny spacecraft had been difficult to identify, which meant the United States still didn't have a firm handle on what they were designed to do. Even powerful Ground-based Electro-Optical Deep Space Surveillance System telescopes had been unable to get decent images of the dark-colored devices.

"Control, Gaspipe's got a tally." Griffin slowly maneuvered the XOV-2 ahead, then above, behind and down behind the nanosats, putting them on his nose. The small satellites were spread across about a mile of space, and seemed to be locked into a geometric pattern. That meant they had some kind of sophisticated station-keeping system on board each vehicle, intercommunicating continuously so they maintained a preset distance apart.

"Control, Gaspipe's in position; taking data."

"Roger, Gaspipe. We're tracking, but you've merged with the targets. You're on your own, I'm afraid," the controller said, his voice tight again. This was a touchy maneuver.

Never before had the XOV-2 been flown to an in-orbit rendezvous with a satellite, and the Lockheed engineers had expressed concern about Speed's ability to control the spaceplane precisely enough to get close, but not too close, to the nanosats. The Pentagon had decided the risk was warranted, though. The intel community had to know what the hell those small spacecraft were doing up here. This entire double-duty mission had been planned to accomplish both the critical hypervelocity, kinetic-energy "Rods-from-God" missile strike against Iran's nuclear facility and a look-see of China's new nanosats. It was a tricky exercise in orbital mechanics, leaving very little margin for something to go wrong.

Griffin made sure the "gun camera," as he'd started thinking of the forward-looking camcorder system, was zoomed properly, recording the entire constellation. He used the handheld Nikon to further document the closest nanosat, snapping multiple images at various zoom and aperture settings, then maneuvered the spaceplane to shoot from other angles. Digital video was essential, but sometimes old-school intel analysts preferred to work with higher-resolution still shots, so he'd get a full complement of 'em.

Little critters are definitely stealthy; dark-colored, recessed lens shields and shaped to deflect, not reflect, radar signals. Clever! He made mental notes, unwilling to transmit his observations via radio, even though the comm link was encrypted.

He was fascinated by the tiny satellites—probably imaging birds, he thought—but time was running out. He had to shoot and run. Onboard "expendables" were marginal at best. In short, he was running out of oxygen and thruster fuel.

"Gaspipe's done, Control. Stowing the . . . uh, 'instrumentation'; preparing to move away."

"Control concurs. You're on the timeline, Speed, but we're cuttin' it close," the ground-based engineer warned. Obviously, the ground crew was monitoring his oxygen and fuel reserves, too.

Suddenly, a brilliant white blast of light exploded in the cockpit. Searing in intensity.

Griffin was partially blinded by the momentary flash of unbelievably bright light. Shielding his eyes with one hand, he tried to find the light's source. Had one of the nanosats fired something at him? Near as he could tell, no. They were all still pointed toward Earth, docile and paying no apparent attention to the XOV-2. Just in case another flash appeared, he closed both eyes and thumbed the mike switch.

"Control! Gaspipe's being illuminated by . . . I don't know what the hell it is! A brief flash of very bright light. Source unknown. Shit! There's another . . . !" A second blast seemed to envelop the XOV, creating a blinding instant of pure blue-tinged light that engulfed the entire spaceplane.

"Gaspipe, disengage! Repeat! Disengage, *now*! Get away from those . . . targets!" the controller said, his voice up an octave.

You bet your sweet ass I'll disengage! Griffin muttered, both hands working quickly, yet carefully. He fired the small maneuvering thrusters to slow down, which dropped the spaceplane into a lower orbit, allowing him to move ahead of the nanosats. He wanted as much distance from them as possible, even though he knew his orbit was elongating, no longer circular. That could be a problem later, he knew. Finally, he was well ahead of the nanosat constellation.

"Control, Gaspipe's clear. Any indication of what that light was? Instrumentation show anything?"

"So far, no. We're looking, Speed. Just get the hell out of there, okay?"

"Gaspipe's well clear; separation's increasing. Setting up for autonav reentry." Griffin glanced outside his canopy, trying to match the terrain far below with his flight plan. *Ah, Christ! I'm over eastern China!* Then it came to him; instantly, he was fairly certain what the flash had been. When he was preparing for his mission, a tidbit of intel generated by a Strategic Command analyst had made an impression and stuck in his brain: China had deployed ground-based laser systems in some of its eastern provinces. Had one of them been fired at the XOV? If so, had it affected any onboard systems? He scanned the instruments, looking for a telltale number out of normal range. Near as he could tell, no. All systems were still in the green and functional.

As he looked outside briefly, something caught his eye, drawing his attention back and to the right. Then he saw it, a tiny web of cracks around the canopy ding, the spot where the Nikon had impacted during its moment of weightless free flight. He stared at the quarter-sized web of tiny lines, his mind racing. *Son of a . . . !* The ding, plus the laser. Could a laser fired from the ground, penetrating miles of atmosphere, have enough residual energy to weaken the thick, yet slightly damaged canopy transparency? He'd have to assume it could.

"Control, Gaspipe may have a problem here," Griffin said. He tried to maintain a nonchalant tone, but doubted he was carrying it off. The aw-shucks Chuck Yeager routine was a little tough for this space-pilot, whose butt was seriously hanging out at the moment. The one-inch-thick canopy was all that separated him from certain death. If it failed during reentry, those beautiful red-and-orange sheets of fire would quickly share the cockpit with him, at least for a few seconds, until he and the XOV-2 disintegrated and burned up.

He summarized the canopy situation via radio, but the ground crew had nothing to offer that countered his best self-advice: get that damned bird back on the ground ASAP. He configured the cockpit for autonav-controlled reentry, then waited until he was over the correct spot above the Pacific. The pitch to tail-forward occurred on schedule, the engines fired, then the XOV somersaulted back to its nose-high position, ready to sink into the upper atmosphere, belly forward. Then it began.

Yeah, he was a shit-hot fighter and test pilot, but he still knew throat-clenching fear. Fear was an old friend that could sharpen the mind, if controlled. Today, he and fear would hold hands as searing heat generated by the friction of Earth's atmosphere engulfed the spaceplane. Fingers of flame, as if guided by an intelligence, would claw at the damaged canopy, trying to get inside the cockpit. If that canopy disintegrated, he was toast. Literally.

He tried to focus on the instruments and displays, ensuring all systems were in the nominal zone. But as purple, then blue, red and orange flames engulfed the spaceplane, his eyes kept flicking to the right. That tiny web of cracks seemed to grow steadily as g-forces and the intensity of those flames increased. Fifty-cent-piece size; then silver dollar. *Damn! How long could it hold?* He was tempted to touch the transparency, just to see how hot it was. But that made no sense; his thick gloves would shield his fingertips from any heat. And the very slight pressure of a touch might be enough to shatter the increasingly fragile canopy.

Just hang in there . . . a little longer; please hold together . . . he half-whispered, half-prayed, coaxing the canopy to remain intact. All he could do was wait. He was gritting his teeth, fighting the punishing, escalating g-forces and buffeting, forcing down a terror that threatened to paralyze his entire being, if he gave in to it.

Then the angry flames were subsiding. He could see blue sky. His spaceplane was through the brutal reentry and was reverting back to its airplane mode. He breathed deeply, gave the ground-control team a quick update, then focused on the overhead, wide, circular landing pattern over California and Nevada.

When the XOV's nose gear settled onto the long runway at Groom Lake minutes later, the sharp bump coincided with an impact streak that shot across the canopy. Griffin swallowed hard. The streak was a massive crack that now spanned a good two feet. If that had happened a few minutes earlier, during the high-dynamic pressure point of reentry . . . He chose to not think about that. He was down and both he and the spaceplane were still in one piece. *Life is damned good!* he declared silently, grinning behind his fishbowl helmet's faceplate.

The XOV skidded to a stop on the runway, surrounded immediately by alien-looking ground crewmen. Wearing protective silver "'scape suits," they quickly checked for deadly hydrazine leaks from the auxiliary power unit, found none, gave Griffin a thumbs-up, and removed their large headpieces. He was cleared to open the canopy, but first waved to draw the crew chief's attention. He pointed to the long canopy crack, then waved the chief to stand clear. He levered the canopy-open switch, leaning to his left. The canopy groaned, but held. Still, Griffin wasted no time clambering out the left side, helped by a husky crewman. The chief found a piece of steel pipe in a pickup truck and jammed it under the canopy, ensuring it would stay upright. Only then did he reach in and retrieve the Nikon camera Speed had stowed in the aft rear compartment.

"Welcome home, sir!" The crew chief smiled, handing Griffin the Nikon. Jamming his helmet under one arm, the chief nodded toward the spaceplane.

"Lucky to get here, I'd say."

"Chief, you have no idea how lucky. I'm damned lucky to be *anywhere!*" the pilot declared, clapping the thick-shouldered chief master sergeant on the back. The pilot fell into a pickup's right seat, then reveled in the refreshing breeze fanning his face as they drove toward the "Site's" control center. *Mission accomplished. Maybe I'm gettin' too damned old for this circus. . . .*

17

AN OLD CHINA HAND

The National Security Council meeting was already in progress when Aster and Lee joined via video teleconference. They had been asked to brief President Boyer about the space-plane mission and the latest wargame findings and to offer much-awaited recommendations. The SecDef was speaking as the link opened.

"Based on our latest recce-sat pass, the strike apparently neutralized Iran's nuclear development and storage site at Parchin. The CIA hasn't confirmed that through humint channels yet, but they're working on it." America had learned the hard way on 9/11 that "technical means" or satellites didn't substitute for old-fashioned, on-the-ground spies. Consequently, "human intelligence" confirmation of overflight data had become standard operating procedure after critical ops.

"That's great news, T.J. Please relay my congratulations and heartfelt appreciation to all involved," Boyer said, then paused, reflecting. Deepening furrows in the man's forehead bespoke what was quickly becoming a chronic state of worry, Hurlburt thought. "And you're absolutely, positively certain the strike can *never* be linked to us, right?"

"Mr. President, the two hypervelocity missiles General

Griffin fired were completely, totally, absolutely converted to energy when they impacted the Parchin facility," Hurlburt said, his palm lightly slapping the conference table to emphasize each term of assurance. "Not a shred of solid metal was left. No trace of explosive—because there wasn't any in either missile—and not one shard of material with 'United States of America' stamped on it. As far as the Iranians know, God himself destroyed Parchin with a bolt from the blue."

Boyer stared hard at his SecDef, then nodded slowly. "All right. I'll take your word for it. But make damned sure NSA vacuums up every snippet of Iranian comm traffic that mentions Parchin. I want to know how that madman president and the Ayatollah react." Boyer leaned back, grasping both arms of his expensive leather chair, and smiled broadly. "One more thing. I still find it hard to understand why *my* Air Force NSC liaison, a general officer, has to fly these spaceplane missions."

The SecDef laughed. "Sir, that was our only option at the time, but we now have another pilot available to fly future missions." Privately, he was certain the Blackstar system wouldn't be available for *any* mission, at least for several weeks. You didn't just go buy XOV canopies at the local Home Depot.

Herbert Stollack, the national intelligence director, spoke up. "Mr. President, I must point out that we still don't know whether *all* of Iran's nuclear weapons have been destroyed. T.J. and I have reason to believe there may still be a couple of Shahab-4 mobile missiles hidden somewhere in the Iranian desert. We're being conservative, though, by assuming they're already armed with nuclear warheads. That possibility needs to be considered as part of any next-step strategy discussions."

"Thanks, Herb. The NSC staff is aware of the CIA and DIA positions about those missiles. Right now, though, I want to address the go-forward recommendations provided by our wargaming teams in Omaha," Boyer said, raising a hand to the wall-mounted screen, now dominated by Aster's and Lee's oversized images. The president frowned and glanced at a single sheet in his other hand.

"But, frankly, T.J., I'm not sure I like *any* of these options." Boyer flipped the page so it spun down the polished conference table toward Hurlburt.

The president was tired. His eyes were sunken, accentuated by a drunk's morning-after under-eye swelling and bagginess, Aster observed. An unshaven five-o'clock shadow added to his disheveled appearance, an uncharacteristic carelessness that had taken White House staffers by surprise. The president was normally meticulous about his physical appearance. But it was Boyer's lack of focus—or was it fear?—that was triggering mental alarms among the assembled cabinet heads.

Was this president capable of directing the potential maelstrom of a global nuclear war? Of comprehending the absolute horror *that* would visit upon the human race? If it became necessary, could he push the button?

Hurlburt had his doubts. This president had campaigned on a position that negotiation, even with terrorists, could solve most geopolitical spats. War was an anachronism, in his view, and a softer, gentler approach would succeed where the previous administration's heavy-handed, aggressive ways had only made matters worse for the nation, he'd claimed. Hurlburt had tried to convince Boyer that such views were decidedly Pollyannish and unrealistic, and would return to bite him *and* America's security posterior, as well. For better or worse, though, a war-weary electorate had bought Boyer's passionate, persuasive campaign rhetoric.

But the current global upheavals, exacerbated by new threats to the U.S. popping up almost daily, had disoriented this naive, idealistic president. His soul-deep view of how things should be, and could be, was suddenly incongruent with actual events. Boyer's "nice-and-negotiate" simply wasn't working.

Welcome to the not-so-nice, non-negotiable real world of snakes and dragons, Mr. President, the self-described "old warhorse" SecDef thought grimly.

"What if we make an appeal before the UN Security Council? And if we do it immediately, won't it show our resolve? That'll put rogue countries like Iran on notice that belligerent sabre-rattling and blatant support of terrorism simply will *not* be tolerated by civilized nations?" Boyer asked, almost pleadingly, looking around the table for assurance and backing. None was apparent, Aster noted. "I'd even be willing to speak to the General Assembly, assuring the global community that we have no intentions of starting a conflict. The world does *not* want a nuclear confrontation. I'll assure the UN that the U.S. doesn't either."

Hurlburt lodged a mild protest. "Mr. President, some of these wargame options are defensive measures that are necessary to protect our citizens and our combat forces. I'd suggest we first consider them, individually if not collectively, before we get into possible UN-related options."

"Yes, yes," Boyer answered impatiently. "But, remember, not every nation views a 'defensive posture' the same way we do. Look at the precedent set by our last administration, and how it was viewed by our European friends!" he said, throwing both arms upward. "Preemptive war, for God's sake! Is that what Iran's Supreme Leader is thinking we're preparing for now? The Great Satan, preparing right now to follow up on Paul's stupid cyber-attack with another 'defensive' move?" The president shot his NSC director a withering glance.

"Can't anybody come up with something more innovative?" an increasingly agitated Boyer demanded, pounding a fist on the tabletop. "Something that will capture the world's imagination *and* stop our own damned hemorrhaging? I feel as if we're spinning—or to use another damned trite, warmonger's phrase I've heard around here lately—*tipping* out of control."

The gravity of the global situation was finally dawning on the president and his NSC staff, Lee observed, staying somewhat detached from the meeting's emotional swings. Those who previously had urged the president to keep all responses focused on terrorism, especially at home, had been silenced, their arguments now moot. Clearly, there was much more at stake, and the nation's leaders finally realized it.

Then Lee's assessment took an immediate setback. NSC chief Vandergrift jumped in, not yet ready to concede a counterterrorism-focused priority. "Sir, we gotta keep the pressure on these terrorists. *They* are the ones killing Americans, right in their hometowns, and that's a lot worse than killing *satellites*," he sneered. "Hell, we've even started replacing some of those space systems. As we speak, there's—"

"Whoa, Paul," Hurlburt interjected. "That's not appropriate for discussion in this forum. Not yet." Hurlburt was incensed, not only by Vandergrift's indiscretion and cavalier approach to operations security but also by the guy's chronic lack of feel for the larger, integrated national security picture.

The full range of the SR-3/XOV spaceplane system's capabilities was still considered top secret, and Vandergrift's mouthing off—even in this top-level meeting—could undermine the president's future options for using Blackstar. The White House staff was filled with self-important, idealistic twenty- and thirty-somethings who were always leaking info to their favorite press-corps reporters. Blackstar's missions would be a particularly tantalizing tidbit for them, he knew. And their disclosure would tip off the Iranians immediately about what had just happened at Parchin.

Further, the Pentagon's deployment of near-space vehicles as interim gap-fillers for lost space systems was being treated as "Special Access Required" information, yet it appeared Vandergrift might have been ready to blabber about *them*, as well! *We sure as hell don't want bad guys shooting at our HARVe platforms already!* Hurlburt fumed silently.

"Well, whatever," Vandergrift dismissed the interruption. "I recommend that Defense immediately deploy its Quick Reaction Forces in support of Homeland Security. Get those folks at Northern Command mobilized; defend the country from within, and then work outward. We need to go into the cities with overwhelming force, block by block. Scare the shit out of anyone who even thinks he can lob a Molotov cocktail through a Wal-Mart's door."

"Preposterous!" bellowed Hurlburt. "We've gotta focus on and prepare for the *external* threat, first and foremost! Let's get real here. The nation is threatened by potential nuclear attack! The International Space Station has been disabled and five astronauts killed, and you wanna wait 'til we hunt down every nut case with a suicide belt full of explosives *before* dealin' with the likes of Iran?" The SecDef was red faced, barely able to control his legendary temper, Aster sensed.

"No, I'm not saying that," Vandergrift countered. "Go ahead and bomb Iran's missiles, if that's what makes your generals happy. But we also need to open negotiations with, or isolate, if necessary, the rest of these international trouble-makers, while we take out terrorists on our own soil."

Admiral Lee wanted to throw up. How many times over his forty-plus years of military and public service had he heard this same lily-livered B.S. about "negotiating" while timely, effective combat operations were hamstrung? How many thousands of brave young Americans had died in vain be-cause politicians insisted on diddling around "negotiating" and "sending messages," only to ultimately have to fight, usually after the enemy was well prepared and dug in?

Yet Stanton Lee himself had been a very effective negotia-tor during his years in the Pentagon, adroitly herding gener-als and admirals who warred with each other over slices of Defense Department budgets. Later, as ambassador to China, he'd come to appreciate how Sun Tzu viewed negotiations as both a strategy and tactic in the art of war. Lee knew that a good negotiator had to know when to hold 'em and know when to fold 'em. But in today's environment, you had to first pick the fight you could win, then see who was left standing before choosing the next battle. And deploying U.S. military units into the streets of cities and small towns, pounding on doors and scaring children half to death, was *not* the war to be fighting at the moment.

Iran notwithstanding, the most serious threat on the hori-zon was China, in Lee's opinion. That one had to be ad-dressed before someone committed a miscalculation that couldn't be undone.

The president jabbed a finger at the two men, and said, "T.J., Paul, you're both right, in a way. We can't ignore those external threats, but I don't have a good feeling about flying Tomahawks or B-2s into Iran, especially on the outside chance those missiles or nukes are *not* there. Remember the

WMDs that were supposed to be in Iraq? I'll be damned if I'll go down *that* road again! On the other hand, we have to get a handle on these brutal suicide bombers. They really think they can bring down the country this way, and we have to prove they can't!"

An uneasy silence settled over the Situation Room. From Omaha, Lee's convictions suddenly crystallized. It was time to act, and time to breach etiquette.

"Mr. President, Stanton Lee here. Pardon my interruption, but may I respectfully make an observation, speaking from a wargaming-adversary perspective?"

His input, obviously unexpected, was greeted with several annoyed glares. Deviating from accepted White House meeting protocol just wasn't done in this administration, which prided itself on reflecting the manners and elegance of Boyer's home state, Virginia. After the Wild West atmosphere of the previous residents, Boyer's crop of genteel Eastern and Southern elitists felt that style had finally been restored to the presidency. That attitude extended to its top-level meetings.

Hurlburt jumped in to bridge the protocol gap. "Admiral Lee is our Red team leader for the strategic wargame under way at STRATCOM, Mr. President."

"Yes, I know all about you, Admiral." Boyer half smiled. "Distinguished ambassador to China, and of course your stellar military career is well known. I'm aware of what you folks are doing in Omaha, too. Damned impressive group out there. Go on, Admiral. I'd be most interested in hearing your Red team's perspective."

Lee bypassed the formalities and spoke rapidly, throwing a Hail-Mary pass of commanding frankness. "Iran is the sword nearest our heart, and that must be dealt with immediately. But swarming Middle Eastern skies with made-in-USA Tomahawks or B-2s to take out missiles and weapons of mass destruction that may or may not exist will trigger an unacceptable backlash throughout the region and in Europe. Even more so than what we saw after the false press reports about Iran's last nuke-missile launch emerged. We've already struck that country's most dangerous threats—those deep underground nuclear facilities at Parchin—and that appears to have been successful. But there's much more that needs to be done."

Vandergrift's patience with Lee was already being stretched. "Admiral, are you suggesting . . . ? Are you saying we should launch an ICBM, or something equally foolhardy?" He clearly did *not* appreciate any implied challenges to his lofty NSC position. Such issues of sweeping national security strategy were *his* territory!

Gilbert Vega, the White House chief of staff, stepped in as the default referee. "Paul, the president wants to hear what Admiral Lee has to say, okay? We can discuss the merits of his proposals when he's through. Please continue, Admiral."

"Thank you. Although I am not prepared to discuss the details of this particular option, I urge you, Mr. President, to consider it as a way to ensure Iran's nuclear, and *only* its nuclear, capabilities are completely neutralized. We cannot accept the possibility of another nuclear attack in that region, sir."

Admiral Lee respected the chain of command and the almost bizarre security protocols that prevented him from discussing classified combat and "deep-black" space systems with the commander in chief. But it still amazed him when people in the same room, who shared the same information, had to pretend they didn't know. Most in attendance probably *did* know about the spaceplane and all it could do, as well as secret U.S. antisatellite systems and the HARVe near-space platforms, but couldn't *admit* they knew. Even in the White House Situation Room! Bizarre. It was as if they were all part of a dysfunctional family and living in denial of the fact.

A hastily scribbled note placed in front of him by SecDef Hurlburt prompted Boyer's casual nod. But did he understand? The SecDef wondered. The note read: TAKE DOWN GALILEO. NO SHAHAB GUIDANCE. The president frowned and glanced again at the note, then back at Hurlburt, but said nothing.

"Mr. President, we also must employ our sea-based missile shields and the prototype airborne laser aircraft," Lee continued. "These are reasonable actions by a nation honor-bound to protect its citizens against a potential attack. And, yes, we can further back that defensive posture by issuing a declaration at the United Nations, as you mentioned: 'No first use of nuclear weapons.' We should then strengthen that declaration by opening NORAD's nuclear threat-warning command center to third-party scrutiny. Let an honest witness attest to what we know: no first-nukes. Canada will concur, I'm certain."

In rich, firm tones, Lee continued to paint bold military and diplomatic strokes, threading innovative concepts through the various elements now contributing to the crisis at hand. Cabinet members and a few staffers paid rapt attention, even though some occasionally pursed their lips and exchanged skeptical glances. Nobody interrupted the silver-haired retired officer, though.

"North Korea must be diplomatically isolated and made to feel militarily impotent. On the diplomatic side, China and Russia must be convinced to lean on North Korea, persuading its leaders to halt all offensive preparations," Lee emphasized. For the past five years the North Koreans had been processing nuclear warheads, then coyly bargaining for time while they activated, shuttered, then reactivated one reactor after another. It was a nuclear-diplomacy shell game, but the world unwittingly played along. North Korea had probably stockpiled at least twenty-five nukes, and all were pressing, no-question-about-it threats. Enhanced No Dong and Taeopdong missiles could put any of those nukes into Tokyo and downtown Los Angeles. Back in 2006 they'd launched a missile toward Hawaii, while wide-eyed Iranians on-site for the event stared in admiration.

"We can expect the North Koreans to denounce or simply ignore a call for restraint, though," Lee predicted. "When they do, we'll have to consider a covert strike against those facilities, just as was done in Iran. 'The Great Leader' will be stunned, but neutralized."

The admiral paused, taking measure of Boyer's reaction. Lee's ideas might appear radical, but years as a warrior and a diplomat had convinced him that a nation's intent must be totally transparent to others. Offer peace, yet remain vigilant. If necessary, strike swiftly against those whose potential to harm is consecrated in ideology and belligerency. *There can be no inconsistency in this philosophy,* he had concluded.

"And what about our streets and cities, Admiral?" asked a subdued, weary president.

"Sir, your leadership and skill in communicating with the American people are powerful tools. I would be honest and straightforward. Tell our citizens about *all* the threats we face—from threats to our security posed by the loss of satellites, to the senseless attack on the space station and the horrific bombing in Denver. Then paint a face of evil on the hateful young men who are schooled in ways that mock the Koran and those of the Muslim faith.

"Be frank," Lee urged. "Suicide bombings are *not* about religious ideology. They're about loss of hope. They're about adolescents who believe there is no future and have nothing to live for. Tell the nation about the loss of dreams faced by many immigrant families—by all our families, in fact. Show our cities' Muslim communities that you understand and you're on their side, yet definitely against the hate and fear radicals among them breed, then nurture."

Stanton Lee's America was a land of second chances and a place where dreams could come true. He still believed that America's institutions—its schools, its industries, its military services—were open doors to anyone. You didn't need to take a number, because you were already welcome. In Stanton Lee's America, a pot of gold waited for everyone. You simply had to have the get-up-and-go to find it. The president must convey such a message. It was the national strength and spirit no suicide bomber could defeat.

"In other words, Mr. President, go on the air and talk about America," Lee suggested. "Invite the very people who disparage us to join us. Our strength is our freedom. Our unity is in our diversity. Talk about America and you cannot fail."

Aster suppressed a smile. Lee was on a roll, but making

damned good sense. For the moment, at least, the old sailor held the NSC and president in the palm of his hand.

"These people have seen violence before, back in their native countries," the admiral continued. "Most came here to find peace. They just want to know which way you're going to lean. Then they'll join you. The fanatics will soon be marginalized. Look at Europe. Anti-Muslim riots still go unchecked by the authorities. Islamic ghettos are under siege because they never assimilated into European societies. We don't have that here in America, not yet. And you can get out in front of the problem. Use your bully pulpit to communicate. At the same time, let NORTHCOM provide assistance as needed, but keep the hard-core soldiers in reserve. We don't need Special Forces roaming our streets. That's not what we do in America. Because if we do, we've lost, and everybody will see we've lost."

"Admiral, you certainly don't pull any punches, but I appreciate that. You've given me much to consider," the president said soberly. "But I'm curious. You haven't addressed the matter of China, except in the context of a North Korean *démarche*."

"China remains a paradox, Mr. President," Lee answered carefully.

"How so? I find those in the opposing party to be a paradox, too. I'm always looking for a way to defeat them." Boyer smiled, evoking a few chuckles.

"Sir, I can't help you with party politics, but I do know a thing or two about China," Lee said, characteristically dry. The admiral's poorly developed sense of humor was legendary. *Resistant to even that of the President of the United States,* Boyer observed with a touch of irritation.

"I'd suggest finding an old China hand, then have him explain the subtleties of Taoism. It is there you may find answers about the enigma that is China." Admiral Lee remembered well the steep learning curve he had climbed to understand his Chinese counterparts. They may be Communists now, but they were Taoists first and always. *Suc-*

cess is passive, he reflected. *Failure is active. Present and future are one.*

"During my time in China, I learned about its people's many-sided views of success," the admiral said. "And at its most elemental level, success is often gained by doing absolutely nothing, just watching. Knowing what one should *not* be doing, and when to not do it—that's the key. The classic Chinese confrontation describes two ancient warriors standing blade-to-blade in the rain, barely touching. One of them moves. He's the one who has lost. He moved without a thought of consequences. That contrasts with the Western instinct to always seek ways to resolve a given situation. You know: shoot first and think later. I don't mean to speak in riddles, Mr. President, but when dealing with China, the more philosophical and the less physical one is, the better an initiative will be accepted. Adversaries are encircled, then embraced, then absorbed."

"Mr. President, *please,*" pleaded Paul Vandergrift, feigning pained exasperation.

"Not now, Paul. Please explain that last statement, Admiral. Straight-forward, please," Boyer ordered.

"Well, the Chinese are masters of subtlety to the point of appearing without structure, even intent. They know the advantages of nonconfrontation, of quiet, and of approaching issues obliquely. They sow uncertainty, which they believe is a way to own an opponent's fate. They can appear to be everywhere at once. Eventually, they absorb their adversaries. Just look at how the Chinese got the British to give them Hong Kong. Then they turned a capitalist pump into their own economic engine. And the people in Hong Kong are more than happy about it." He paused to take a measure of the president. "That's a classic Chinese strategy. Look how effective it is."

The president sat quiet for a moment, mulling over Lee's truisms. Then, for the first time in weeks, he indulged in his trademark smile. "Thank you, Admiral." Boyer folded his hands and leaned forward, elbows on the table, indicating a shift in the meeting's tone and direction. "General Aster, we will review your recommendations in light of what we've heard here. Please continue your deliberations in Omaha, and I'll be back to draw on your counsel if I need additional input. Thank you, gentlemen."

As the link to STRATCOM went dead, Boyer turned to the Cabinet members, lips set in a tight, thin line. "I want State and Defense to work up details implicit in each of the admiral's international recommendations. Homeland Security, you will vigorously pursue whatever actions are needed to deny any terrorist a spectacular suicide-bombing target. And I want the Attorney General's Office to give me a prioritized list of my emergency powers on the domestic front."

Boyer turned to his longtime friend and chief of staff. "Gil, schedule a press conference within the hour. Tell the press corps it's urgent—national security urgent. And somebody, anybody, find me a China hand like the admiral described. If you can't, get Lee out of Omaha and bring him to Washington ASAP."

18

GALILEO

The group of small tactical satellites that Speed Griffin and the XOV-2 spaceplane had only recently released into orbit drifted silently in the black of space, hard at work. At the moment, each tacsat's sophisticated optical system was staring at a particular spot deep in Iran's desolate outback. But individual nanosats focused on slightly different patches of sand. Via laser communications links, digitized images were fed from those nanosats—all flying in extended formation, each within a few miles of, and a precise distance from, its neighbors—to the "mother hen" nanosat located at their center. The latter was separated from the other tiny spacecraft by equal distances, positioned as the hub of a large wheel.

"Mama Hen," as its engineer-designers referred to the central platform, appeared outwardly to be a duplicate of its nearby nanosat "chicks." Unlike the chicks, though, Mama Hen also carried a diminutive computer among its compact electronic payload components. Programmed as a "server/router," the powerful microprocessor digitally stitched each of its chicks' slightly overlapping images into a larger, single image, then routed the compiled bits of data to a covert ultrawideband transmitter aimed at ground stations on Earth. New photos were fired to those stations several times a minute, then flashed across the globe, usually via communication or classified relay satellites, to CIA headquarters in Langley, Virginia.

The downlinked images were no ordinary family-album-variety photos. Each was a "hyperspectral" treasure-trove of data that, when computer-processed and false-colored, revealed a wealth of information that normally would escape the human eye. The magic of hyperspectral imaging emerged by literally slicing the visible-light spectrum into thousands of tiny bands. Each band of light could reveal specific information about the target.

For example, when a hyperspectral sensor flew over a particular region of the globe, a few quick taps of a workstation's keys could extract only those bands of light that were reflected by a certain type of fish in the Atlantic, or a few mahogany trees in a Brazilian rain forest. Looking at a standard photograph of the ocean or the Amazon, one only saw water or a blanket of trees. But tell the computer to highlight those few wavelengths of light reflected from a school of tuna just below the ocean's surface, or bounced from a mahogany tree's foliage, and the image was magically transformed, revealing those specific features. The information was already there, buried in the image, but hyperspectral technology allowed it to be plucked from homogenous backgrounds, then highlighted through computer-generated false coloring.

At the moment, the nanosat "chicks" were simply vacuuming every wavelength of light being reflected from Iran's badlands. But Mama Hen was looking for a few specific wavelengths, obeying uplinked orders from a nondescript antenna in Oman. At CIA headquarters, hyperspectral-smart technicians told their powerful workstations to further enhance any of Mama Hen's images that had certain characteristics. Several showed some very interesting features, a specialist decided.

Moments later, those images were being closely examined by the CIA's director of operations. Another few minutes and the agency director had seen them. He, in turn, flashed them to the national intelligence director. The NID was both ecstatic and scared out of his wits. When these images were matched with information he'd received only hours before from the

Defense Intelligence Agency, the military's primary intel-gathering arm, there was little doubt remaining: Iran was preparing to launch another Shahab-4. The NID grabbed a phone and called the secretary of defense. This was shaping up to be another long night.

23 APRIL (0235 LOCAL TIME)/WHITE HOUSE SITUATION ROOM

"How can you be so damned sure, Herb?" President Boyer growled. "Sounds like a lot of circumstantial evidence that would never stand up in a court of law. All pieced together so it appears to be something it isn't." Boyer, awakened by a red-phone call to his White House bedroom, was trying to make sense of high-resolution images projected on a large wall-sized screen in the almost empty Situation Room. He was dressed in a pair of hastily donned gray sweatpants and a matching University of Virginia sweatshirt. White socks and moccasins completed his casual, middle-of-the-night ensemble. Uncombed hair, strands aimed in random directions, matched the man's grumpy mood. Like his predecessor, he valued a good night's sleep, and didn't take kindly to having it interrupted, even by an urgent call from Herb Stollack, the national intelligence director.

"Mr. President, I'll skip the technicalities, but trust me: We're damned sure we know what we're looking at here," responded Stollack, his own impatience showing. "These images confirm that the Iranians are definitely preparing another Shahab-4 missile for launch. They haven't started the final fueling process, but based on activity at a known launch site—that's the image you're looking at now—they've already mounted a warhead on that missile. They wouldn't do that unless . . ."

"Okay, you've made your point. And I suppose you're right. We have no choice but to assume it's another nuclear warhead," Boyer snapped. "But what's this B.S. about Galileo receivers? Maybe I'm just not awake, but that connection escapes me."

SecDef Hurlburt's voice appeared. "Herb, if I may . . . ?"

"All yours, T.J.," Stollack clipped.

"Mr. President, NSA picked up a flurry of communications traffic between a French electronics company and a known Iranian Revolutionary Guard missile expert in the last few days. Based on NSA's tip, some of our DIA folks were able to confirm the shipment of several Galileo satellite navigation receivers from the French company to Iran. At least four receivers—and they're all top-end, sophisticated types, only suitable for something like a guided missile or high-speed aircraft."

"So, this has something to do with that note you slipped me yesterday, during the NSC meeting? It mentioned 'Galileo,' but . . ." Boyer mumbled, his words trailing off.

"Yes, sir. We already had a strong suspicion that those Galileo receivers were slated for installation on an Iranian Shahab or a cruise missile. Only thing that made sense," Hurlburt concluded.

"What does this all add up to?" Boyer asked. He was afraid he already knew, but wanted it spelled out.

"It means Iran's next missile shot won't miss," Stollack said bluntly. "We suspect they figured out why that last Sha-hab missed its target and landed in the drink. Its guidance system relied almost exclusively on GPS signals, which we know were severely degraded in that part of the world at that particular time, thanks to the maser kills on four of our birds. Those damaged GPS satellites caused the missile to miss its target; thank God for the one bit of good that prevented a dis-aster. But it looks like clever Iranian engineers figured out they only need to switch over to Galileo. Then they can use nav signals from the European version of GPS for guidance *and* selectively exclude our GPS data."

"I thought the Galileo constellation was pretty much worthless right now," Boyer said, surprising Hurlburt. The president may be a defense lightweight, but the guy was a quick study, and had an awesome memory.

"To a degree," Hurlburt interjected. "The Euros only have three front-line satellites in orbit right now, because they just started populating the constellation with production platforms.

Originally, any commercial ground or airborne receiver was supposed to be able to use both Galileo and GPS signals. But the Euros decided about three years ago to forget compatibility and go with their own signal format.

"There are still a few Galileo prototypes up there, too, but we know they're starting to degrade and aren't reliable," the SecDef continued. "That's why we really hadn't been concerned about Galileo being used for missile guidance and targeting. That is, until NSA and DIA got wind of those French receivers going to Iran. They're brand-new, and some of the intercepts made it clear that these units could ignore the prototype's signals."

"And only pay attention to the three production ones?" Boyer asked.

"Bingo, sir. And that's what got our attention," Stollack said. "Once those receivers are integrated onto a Shahab-4, the missile becomes fairly accurate, but only at certain times. It has to have all three Galileos at exactly the right orientation in the sky to ensure a good navigation solution."

"Can we predict when that will occur?" Boyer asked.

"Yes, sir. And we've done that," Hurlburt said. "The next proper conjunction of three Galileos happens day after tomorrow."

"What are our options? Did the Omaha wargamers consider this possibility?" Boyer asked wearily. Again, he already knew the answer but wanted to hear it rather than assume it.

"Yes, they did," Stollack confirmed. "The possibility of Galileos being used by an aggressor for an attack in the Central Command Area of Responsibility was postulated by Admiral Lee's Red team. T.J., over to you for the options."

"The 'gamers came up with five mitigating measures, sir," Hurlburt continued. "One, take the diplomatic route. Try to get the Euros to temporarily shut off their Galileos, thus denying the Iranians any reliable space-based guidance capabilities. Two, take out the Shahab launch sites with an air or spaceplane strike. Three, deploy missile-defense capabilities to the region. Four, take out at least two of the Galileo navsats. And, finally, attempt, through back-channel diplomatic routes, to convince

Iran's more moderate elements to force the radical elements to stand down those missiles."

"I suspect the last one's out, since we have few or no effective back channels to Iran anymore," Boyer groused. "Paul Vandergrift saw to that, and got that good-lookin' Adkins woman killed in the process! Christ . . . ! What about one through four?"

Hurlburt took a deep breath. "None are great options, sir. I had some of my people approach State, before yesterday's NSC meeting, asking them to work on a back channel to Iran. They said that was absolutely not an option anymore, as you said. So, we asked State to approach the Euros about exercising navigation 'shutter-control'—turning off the Galileo birds. At least taking them off-line long enough to buy us time to work these other options. Bottom line is . . . well, to be blunt, sir, we were pretty much told to 'F' off."

"Why am I not surprised?" Boyer grimaced, running a hand through his hair in exasperation.

"The French took great offense at the idea one of their companies had bootlegged some advanced Galileo receivers into Iran," Hurlburt added. "They huffed up, got all indignant, and that killed what little Euro Union willingness there was to shut the Galileo birds down, even temporarily. Officially, they told State that there were 'already too many users depending on Galileo.' As you may know, Galileo was sold to all of Europe by promising it would never be shuttered by military pressure, unlike GPS and our shutter-control policies. Turning the birds off, even temporarily, would cause 'adverse impacts on critical users,' they claimed. So, the diplomatic route is a nonstarter, sir." *And please don't even think about wasting time trying again,* Hurlburt thought but didn't say.

Boyer surprised him a second time. "You're right, T.J. The damned Euros would rather take a Shahab in the chops than cut into a lucrative revenue stream. They proved that European mercenary tendencies will always trump global security prior to the 2003 war in Iraq. But, with only three satellites up, how can anybody be reliant on Galileo already?"

Stollack stepped in again. "That's correct, Mr. President. Our tech analysts say there're only a few times each week when all three of those platforms are in the right locations to enable the necessary triangulation for good nav accuracy, at least from the perspective of anything launched from Iran. And the bulk of receivers now fielded are dedicated to testing the system while it's being populated. No user of any consequence is relying on Galileo for actual navigation yet. It's way too spotty and intermittent. Somebody might be using the timing signal now and then, but that user base appears to be mighty small, as well. Most are still dependent on GPS, such as it is at the moment."

"Okay. That leaves three options. And I fail to see how any of those warrant a discussion at . . . three A.M.," Boyer said pointedly, glancing at a wall clock. He was still groggy, and dawn was less than three hours away. "Exactly why are we having this conversation *now?*"

There was a long pause. Neither Stollack nor Hurlburt was anxious to take the next step. Hurlburt bit the bullet first. "Sir, I've already ordered a couple of modified Aegis missile-defense ships to the Mediterranean. They'll be our first-line guard against the possibility of another Shahab launch. But they won't be in position for a couple of days, at best. And our one and only Airborne Laser platform—a Boeing 747 freighter with a humongous missile-killer laser onboard—is temporarily down with a maintenance problem. It's at Edwards Air Force Base now, and we can't get it into the Med in time to counter a Shahab."

"Jesus! I thought ABL was still years away from being operational! Why the hell is it even being considered?" Boyer exploded.

Hurlburt had expected as much. "Absolutely right, Mr. President. ABL *is* still in development. But it's already demonstrated it can shoot down pressurized-body missiles like the Shahab-4. It did that two years ago, even though the critics said ABL was a waste of money and couldn't possibly do it," the SecDef added, unable to resist the jab. One of those critics was Boyer himself, who had expounded repeatedly against the ABL program during the 2008 campaign—defying the recommendations of his primary defense advisor, T. J. Hurlburt, of course. "You heard Lee say we need a missile shield and we need it now. A half-ready ABL is better than nothing, and we may need it before this nightmare is over."

"Spare me the I-told-you-sos at three friggin' in the morning, T.J.," Boyer growled. "Get to the damned point. What do we need to do and when?"

"Air strikes are out. We covered the in-atmosphere conventional reasons before the Parchin nuke facility strike: too provocative, hardly covert, and we'd get killed by the Euro and Mideast media and anti-Americanists. An overt attack would further inflame the radical Islamists *and* send the rest of the Muslim world back into the streets. Plausible deniability is mandatory.

"We have no choice, sir," Hurlburt rushed ahead, unable to resist smiling to himself. *Score one for the real-world team!* "Right now, I want your authorization to take out two Galileo satellites."

"Why not hit those missile sites with the spaceplane again? Use those Rods-from-God . . . ," Boyer began.

"The XOV-2's down, sir. We're trying to get another canopy for it, but that'll take awhile. For the near term, it's out of commission and not an option." *And that tidbit was in your intel briefing yesterday morning,* Hurlburt didn't say.

Boyer left a long silence hanging in tele-space. Finally, "Yeah, I recall now. Hitting Galileo's our only option? If we don't—?"

"I'm afraid so, sir," Stollack interrupted. He and Hurlburt had been through this scenario several hours ago, tag-team grilling their experts about every angle possible. "If we don't take out a couple of the Galileo birds, we have to assume any Shahabs that are launched will have reliable space-based guidance, and they *will* hit their targets. And they'll have their first launch window in roughly thirty-six hours. We won't have a missile-intercept capability in place by then."

"And Iran's probable targets, Herb?" Boyer asked, obviously examining all relevant ramifications.

"Israel's the top pick, sir," Stollack said tightly. "The Israelis have their Patriots and Arrow missile-defense batteries locked and loaded, so they're as ready as they can possibly be. And, if Iran launches at Israel you can imagine the response. The whole Middle East will be in flames within hours, and before the nuclear dust settles, the rest of the world will be without oil."

Hulbert took the cue while Stollack's vision of Armageddon settled over the teleconference. "Also, any of our bases in the Gulf and Central Asian regions are vulnerable. And, as we saw the last time, pretty much all of southern Europe's within Shahab-4 range."

"Christamighty! That puts several million people at risk! The . . . the damned Euros would rather play piss-up, cross-Atlantic, we'll-show-you politics, and *assume* those hothead Iranian fanatics won't lob another nuke in their lap, than take precautions and give their citizens a fighting chance! Incredible. Staggers the imagination." Boyer paced around the Situation Room, relying on a speakerphone to snag and transmit his ramblings. He was vaguely conscious of his always-present Secret Service shadow hovering nearby, pretending not to listen.

"And we simply can't allow the Israelis to launch their own preemptive strike against Iran. The entire Mideast *would* explode. So, we're left with no choice but to take down those Galileo navsats, right?" The president was struggling, trying to talk himself into accepting the only solution left on the table. But he obviously wasn't ready to go there yet, Hurlburt observed, saying nothing.

"T.J., there's obviously a time factor here that you haven't mentioned. Enlighten me," Boyer ordered.

"That's correct, sir. With the spaceplane out of commission, we've only been able to come up with one option that's available in the timeframe we have. That's two F-15 fighters firing some big-assed antisatellite missiles from high altitude, taking out two Galileo birds. We have two Eagles being outfitted at Edwards now. Our apologies for waking you up, but the only near-term window for an ASAT mission is shortly after daylight today, California time. That's when two of those Galileos will be in the right place, giving us a shot at them. We won't have another opportunity for a couple more days, and that could be too late."

"Just a few hours from now? How the hell did you get missiles like that rounded up so fast?" Boyer asked suspiciously. "I sure as hell didn't authorize any ASAT program!"

"Correct, sir. But your predecessor did. It was a Special-Access-Program authorization, so very few people were read-in. But it's a legal, standing presidential directive," Hurlburt explained. "He ordered the air-launch option be resurrected, because it was the only near-term, proven, in-orbit counterspace capability we could field in short order. Only one ASAT-missile shot like this has ever been done, and that was back in 1985, but it was still the best option for our space forces in the near term."

"The United States actually shot down a satellite? With a missile . . . from a fighter?" Boyer asked, incredulous.

"Yessir. It was strictly a test, and the target was one of our own defunct spacecraft. But the missile hit that satellite dead-on. It was a kill, pure and simple."

"I'll be damned. . . . How many of these ASATs do we have on hand?"

Hurlburt hesitated, scanning the "are-you-cleared-for-this?" part of his brain, an unconscious screening technique born of many years spent balancing classified and unclassified information. *This is POTUS, for God's sake!* he chastised himself in a flash. "Not many, sir. But enough for our immediate needs. We've ordered contractors to pull out all the stops and build more of 'em ASAP. Who knows how many ASATs we'll need before we work our way out of this mess? And right now, that's the only real ASAT capability we have. After that Chinese laser attack on Speed's spaceplane, I'm not about to risk using the XOV in that role. Besides, Blackstar can't get high enough to take out a Galileo."

"What's the probability of success, assuming I decide to target those Galileos?" Boyer asked, still probing, still not ready to green-light an ASAT mission.

"About seventy to eighty percent, sir. Those birds are in a much higher orbit than the spacecraft shot down in '85 was," the SecDef continued. "Our lab rats and test folks claim the new ASAT missile can reach mid-Earth orbit, but it's never been done before. Frankly, to me, it's like using a .22 rifle to shoot a fast-pitch baseball out of the air. Possible, but damned tricky. Of course, we have terminal-guidance sensors on the missile, and they'll help guide the kill-stage through the final intercept." Hurlburt hoped he sounded more confident than he felt. Assurances from techies always made him nervous. He'd seen too many failures when so-called slam-dunk systems had to actually perform in real, not virtual, environments.

Boyer continued to pace around the table, repeatedly running a hand through graying night-wild hair. He finally leaned on the conference table and spoke into the speakerphone. "Gentlemen, I'm clearing you to go ahead with the ASAT mission. Do whatever's necessary to make damned sure those Shahabs can't get a good guidance solution. If I understand this correctly, the Iranians probably wouldn't even attempt a launch once they see Galileo is a lost cause, right?"

"Good point, sir. I'd think a loosey-goosey Galileo signal would be reason enough to *not* launch," Hurlburt said, impressed with his president's line of logic.

"But I want the option of calling off those ASAT shots right up to the last few seconds before trigger-pull, understand?" Boyer didn't wait for acknowledgment. It was an order from the president, after all. "Herb, I want your people to scrub all pertinent intel about those Shahab-4 preparations and give me a last-minute assessment of launch probability, okay?"

Stollack agreed, jotting a few notes to himself. Hurlburt also weighed in with a curt "Understand, sir."

"I don't have to tell you how serious this step is, gents," Boyer said, much quieter. "Unless this old Virginia lawyer is totally wrong, I'd say we're about to commit an act of war against traditional allies. We damn sure better not screw this up. And, once again, 'plausible deniability' is a given; you know that. *Damn!* Seems like everything we do these days has to be a 'black' op!" The open line was silent. Without a pleasantry of signoff, the president punched the speakerphone's disconnect button.

Walking back to his White House bedroom, head bowed in thought, Boyer allowed himself to acknowledge how bone-deep weary he felt. At the door, he turned and nodded to the USAF officer following at a discreet distance, carrying what could only be *the* aluminum briefcase. The codes inside that case could unleash a horde of American ICBMs, bombers, cruise missiles, and submarine-launched thermo-nuclear weapons. And it was his, and only his, decision to do so . . . or not.

The president shivered, then closed the bedroom door.

19

ASAT

Major Bret "Roach" Rochelle increased pressure on the toe brakes, carefully bringing his F-15C Eagle to a halt, per the crew chief's direction: crossed wrists, arms raised above his head. Once wheel chocks were in place, Roach placed gloved palms on the fighter's glare shield, elbows resting on each canopy rail.

That unambiguous act, showing his hands to those on the ground, assured the crew chief that the test pilot wasn't messing with cockpit switches. Simple, but it had become a standard procedure to ensure the safety of a "Last Chance"–area ground crew working under an aircraft. That crew chief ducked under the fuselage, inspecting panels, looking for fuel or hydraulic leaks, generally making sure all safety pins had been pulled and the fighter was ready to fly.

Another uniformed two-man crew, the armament team, also disappeared under the aircraft, duckwalking beside a long telephone pole–like missile attached to the F-15's centerline weapons station. One weapons specialist pulled the last red REMOVE BEFORE FLIGHT safety streamer and ducked under the left wing. He held a handful of various-sized streamers aloft, fanned so the pilot could count pins attached to four canvas ribbons. *Celestial Eagle II*, a name freshly painted on the fighter's fuselage, was good to go.

Rochelle gave the crew chief and airmen a thumbs-up, then turned to an open checklist strapped to his left leg. On the other thigh, a mission test card was pinned under a plastic clip sewn into his olive-drab g-suit. With the comfortable familiarity of more than a thousand hours of Eagle flight time, the officer quickly completed pre-takeoff checks, configuring switches and pressing buttons that rimmed various flat-panel multifunction displays on the fighter's instrument panel and side consoles. He then ran through a personal mental checklist by double-checking his oxygen system connections, ensuring the O_2 eyelid blinked white, then dark, as he breathed in and out. *Seat-arming to go. Get that as we take the active,* he flashed.

The flight leader glanced to his right, where his wingman, Major Steve "Sierra" Hilton, was head-down, doing the same in an identically configured jet. "Eagle flight, radio check," Rochelle radioed.

Hilton glanced up, his face hidden by a dark visor and oxygen mask. "Eagle 2's got a problem. Stand by."

Rochelle double-thumbed a transmitter switch on the twin throttles, acknowledging. *Ahh, crap! Now what?* He grimaced, stomach tightening. Although an abundance of contingency time had been built into the mission plan, he hated to be eating into that reserve already, before takeoff. He and Hilton absolutely had to have their fighters at specific points in the sky at very precise moments, or the mission would be a bust. And the ramifications of missing the time/position window were horrendous. Borrowing a well-known quote from NASA's Apollo moon-shot program, the two pilots had been told "Failure is not an option" by a grim-faced Air Force Flight Test Center (AFFTC) commander during their mission briefing.

"Eagle 1, 2 here. Go button four," Hilton clipped.

Not good! Rochelle one-fingered the upfront control panel, switching to a dedicated fighter combined test force frequency allocated to this mission.

"Eagle 1's up. What's the problem, Sierra?"

"Nav system's acting up. Inertial's okay, but the GPS primary's intermittent," Hilton replied. Rochelle mentally scanned the Eagle's navigation system. The inertial-aided GPS box was usually rock solid. Then he remembered. *Damn! The new OFP!*

A new Operational Flight Program, basically updated software, had been installed in all of Edwards Air Force Base's test aircraft to selectively screen out nav signals from four maser-zapped GPS satellites that were spewing bad data. Of course, his and Hilton's aging F-15s had been far down on the list, because the old Eagle was in the to-be-retired column of the USAF fighter inventory. These two test jets had been updated with the new OFP only yesterday, one of many last-minute preparations for today's shot.

"Test, you copy?" Rochelle asked casually, knowing he was on "hot-mike." Everything he said in the cockpit was telemetered to a control room over at the Ridley center. He didn't have to transmit over the UHF radio to be heard.

"Rog, Eagle 1. We're taking a look now."

"Could it be the new software? The new OFP?" Rochelle asked.

"We think so. I've got maintenance scrounging for a replacement unit," the flight test engineer said.

Rochelle's gut tightened another notch. Besides being one of the fighter CTF's test pilots, he doubled as its operations officer. He knew there was only one other Eagle on the flightline with that new GPS software installed, and it was scheduled for a classified-system evaluation flight this afternoon.

In Edwards's Ridley control room, General Howard Aster ripped a headset from his ears and turned to an officer seated next to him. "Lance, do these people fully understand what's at stake here? What's riding on this mission?" His tone was more demanding than brusque.

Brigadier General Lance Ferris, the AFFTC commander, was already tight-jawed about the STRATCOM chief's sudden appearance early this morning. Ferris had just crawled out of

ASAT   335

bed when the command post called his private quarters to report that Aster was inbound. The four-star's C-21 Learjet had been taxiing up to the Edwards base operations building when Ferris arrived, without breakfast or coffee, of course.

"Sir, believe me. They know. I went over everything, in detail, with these pilots and the entire test team last night, right after your people gave us the heads-up." Ferris held Aster's stare. The tall STRATCOM chief's hard blue eyes were red-rimmed and bloodshot. Obviously, he hadn't slept much, either.

Aster half smiled and nodded. "Good. It was a short night in Omaha," he quipped, a roundabout apology for his outburst. He turned and sipped the oversized mug of hot coffee he'd been handed when he arrived at the Ridley mission control center an hour earlier. Surveying the room, he noted two short rows of men and women manning sophisticated workstations, each a virtual mission control center in itself. A large screen at the front of the room displayed a wavy, heat-distorted TV image of the two F-15s now idling near the east end of Runway 22, a 15,000-foot stretch of concrete on the edge of Edwards's famed Rogers Dry Lake. Long-range cameras from a tracking and imaging facility atop a windswept hill overlooking the stark Mojave Desert airbase were trained on the Eagles, beaming real-time video to the test team's control room.

Aster nodded toward a medium-built man with thinning hair. Dressed casually in slacks and an open-collared golf shirt, he had just entered by a side door. "Is that Pearson?" he asked Ferris.

"Yessir. Retired Major General Doug Pearson. He had my job back in the late nineties, early 2000s."

"And the only guy to ever shoot down a satellite," Aster added, watching as the former test pilot was greeted by several engineers, each standing to shake hands as Pearson worked his way down the line of workstations. "Seems to be well liked," Aster observed.

"Doug spent a lot of time here at Edwards. Had tours as a young test pilot, then he came back and ran the place for about four years," Ferris explained. Before Aster could ask about such a long tour, unusual for a flag officer, Pearson approached, hand extended.

Ferris introduced the two men, then listened as Aster spent the next few minutes learning about the retired two-star's remarkable 1985 test flight. Then-Major Pearson had fired an LTV/Boeing-built ASM-135A air-launched antisatellite missile from an F-15, unofficially designated "Celestial Eagle," on September 13, 1985, destroying P78-1, an American space-test-program satellite orbiting 286 miles above the Earth. For years, Pearson had been called the only "space ace" in the Air Force.

Aster stole a glance at his watch, then at the big screen. Neither F-15 had moved, their engines' hot exhausts distorting light reflected from the sun-baked clay lakebed behind them. A needle-nosed F-16D, painted in the white-and-orange color scheme that identified Fighting Falcons used primarily by the USAF Test Pilot School (TPS), turned into the quick-check area, partially blocking the test team's TV image of the ASAT missile–armed Eagles.

In that F-16, a French officer, an exchange TPS student, studied the single long missile mounted under each F-15, careful to keep both hands visible as a crew chief scurried under his jet. The missiles were about six meters long, roughly 50 centimeters in diameter, he estimated.

"What ees zat on zee F-15's centerline?" he asked, his English laced with a heavy accent.

"Beats the hell out of me. Never seen it before," answered his flight test engineer, another TPS student, from the rear cockpit. He, too, studied the long, thick missile. "Some kinda rocket with several stages and a nozzle. Sure is a big mutha!"

The crew chief gave a thumbs-up, clearing the F-16 to taxi the last few hundred feet to Runway 22. The French pilot turned back to his cockpit, still curious about the missile. There were many unusual projects here at the USAF's premier development test and evaluation base, he knew. Most were cloaked in secrecy, and even more were "NoForn," meaning non-Americans need not ask. No matter. He'd find somebody who'd talk about the huge "blivets" under those Eagles. He looked right, then left, released the toe brakes, and let his F-16 creep forward. He had a mission to fly. And TPS students were too damned busy to worry about anything that didn't count toward graduation.

The French student would never know the far-reaching impact of his subsequent innocuous decision, made later that day, to *not* send an e-mail message to his commander in France. Had he relayed his observations about the two missile-armed F-15s, he might have triggered yet another ugly international political tiff between France and the U.S. The almost-confrontation was avoided by a simple fact of TPS-student life: studying for a big test the next day, he was too busy to send an e-mail. Besides, he was just an air force captain. Who'd care that he'd seen huge missiles under a couple of Eagles? Probably just laugh it off anyway.

In the lead F-15, still in the "Last Chance" area, Rochelle was starting to sweat the proverbial bullets. Normally cool and calm—one had to keep the stereotypical test-pilot image alive, after all—his mind raced through every contingency plan already briefed, then jumped to a few admittedly wild-assed ideas that hadn't been considered. One way or another, he needed to get two ASAT-armed Eagles in the air, and damned soon. A very small window of opportunity was fast approaching.

He noted a contractor's pickup truck swinging into the quick-check pad, smartly braking to a stop well away from Hilton's F-15. A civilian jumped out and raced toward the jet's crew chief. Wearing headphone-like ear protectors and carrying a gold-tinted avionics box trailing a short, thick pigtail of wires and circular screw-on connector, the man shouted something into the crew chief's headset, eliciting a quick nod. The sergeant motioned to an airman, directing him to retrieve a metal ladder lying nearby, and mumbled something behind his mouth-mask. The pilot chopped his fighter's right throttle, shutting down the starboard engine.

Hilton thumbed the canopy switch and saw the crew ladder

being hooked over the right side rail while the big bubble was still rising. He dropped his oxygen mask and yelled as the civilian's head appeared.

"What's the story, Mac?"

"Gotta replace your IMU controller, Major," Mac yelled to be heard over the Eagle's screaming left engine, earsplittingly loud, even though it was idling. Hilton's flight helmet made it that much harder to hear, too.

"Where'd you get that?" Hilton yelled, pointing to the avionics box. Like Rochelle, he was sure there was only one other reprogrammed F-15 inertial measurement unit in the CTF, and it was on that Eagle slated for a high-priority mission later in the day.

Mac gave him a raised-eyebrow glance, while leaning over the canopy rail, screwdriver in hand. He went to work immediately, expertly loosening fasteners on the right-hand console, but yelled back. "Don't ask, sir. You don't want to know!"

Hilton grinned. The high-priority F-15 had just been cannibalized. *Ya gotta love resourceful maintenance troops!* Get *your* bird in the air and take the lumps later. "Good on ya, Mac!" Hilton yelled. "Easier to beg forgiveness than get permission, eh?" Mac shot the pilot a wink and kept working.

Back in the Ridley control center, Aster was losing patience again. "Damn! We've gotta get those jets off—*now!*" he barked. Frustrated, he turned to Ferris. "Lance, I want to talk to that Eagle flight lead; see what the hell's the hold up! Patch me through to him!"

Ferris blanched, but nodded tentatively and moved toward the flight test engineer serving as today's test conductor, the only person in the room allowed to talk directly to Rochelle and Hilton. Although a dozen or so engineers and specialists monitored real-time telemetry data from the two F-15s, all communications were routed through a single person, the test conductor.

"Excuse me, General." Aster turned to Pearson, who was standing at his right elbow. "If I may . . ."

"Sure, Doug." Aster nodded, still tight-jawed. *How can he be so damned calm?* the four-star wondered. *Must be nice to be retired, out of the line of fire.*

"Just an observation," Pearson began, speaking softly, a slight drawl accentuating his natural aura of cool and calm. He flicked a hand toward the big screen, where the F-15s still waited, engines gulping fuel. "Nobody wants to get those fighters in the air more than the two pilots and that ground crew. Believe me, they're doing everything in their power to get those Eagles moving. And I'm not sure a call from you right now would really be of much help, sir."

A trifle stunned, Aster, hands on both hips, turned and faced the shorter man. Looking along his left shoulder, Pearson met Aster's glare, and casually crossed his arms. Neither man's gaze wavered for a long moment, two warriors standing eye to eye.

Aster smiled slowly and laughed, softly at first, then heartily. Shaking his head, he lightly punched Pearson's upper left arm. "Doug Pearson, you're *so* right. Damn! I needed that!" He laughed again, attracting a few puzzled stares, including one from Ferris. Still grinning, Aster caught Ferris's eye. The STRATCOM chief drew his fingertips back and forth across his own throat, while mouthing "Forget it!" Ferris nodded and said something to the test conductor, who shrugged and turned back to his screen.

In the "Last Chance" quick-check area, Mac gave Hilton a thumbs-up. The nav system was working now, a conclusion the pilot had already come to, as well. He lowered the canopy, made sure it was locked, received an all-clear from the crew chief, and restarted the right engine. Hilton then called his flight leader as the crew chief yanked the wheel chocks clear.

"Eagle 2's good to go." Rochelle nodded, exhaled in relief, and nudged two throttles, following his crew chief's taxi-forward direction. Another nose-dip indicated a final brake check, then the jet angled toward the main runway.

"Tower, Eagle flight's number one for takeoff," he said.

"Eagle flight, cleared for takeoff," the tower controller

answered immediately. He, too, had been briefed to give the pair of F-15s whatever they wanted, whenever they wanted it.

Seconds later, Rochelle pushed both throttles forward, scanned the hydraulic, fuel, engine, electrical, and other gauges, released his toe brakes, then jammed the throttle levers forward, past a detent, commanding full afterburner. The big twin-tailed fighter leaped forward, quickly gaining speed. Rochelle waited, holding the jet on the ground a bit longer than he normally would. The 2,600-pound, blunt-nosed missile slung under the Eagle's fuselage demanded more airspeed to create enough lift to compensate for both the extra weight and drag. A firm, steady aft stick movement rotated the nose up and the jet was airborne. Rochelle mentally confirmed that the fighter was indeed climbing and still accelerating, then raised the gear handle, tucking the main and nose gear into their fuselage wheel wells.

He started a wide, climbing turn to the right and pulled the throttle out of afterburner. A slight jolt of deceleration, actually, loss of acceleration, answered as the 'burner snuffed out. Momentarily, Hilton's jet slid into a loose right-wing position.

"Eagle flight's airborne," Rochelle said, prompting a second controller's clearance to switch frequencies. "Eagle flight, button four." Two clicks said Hilton had copied and was switching to test frequency.

Over the next few minutes, the two fighters climbed to altitude and turned west, heading for the Pacific Ocean. Ground-based towers relayed their telemetered hot-mike comments and data streams to the test center at Edwards, keeping the team's airborne and ground-based members in constant contact.

Aster said a silent thank-you prayer as the two fighters slowly disappeared into the big screen. *Still a long ways to go, but at least they're airborne!*

"Kinda makes you wish you were up there instead of down here, doesn't it?" Pearson mused, level-toned. He and Aster were both fighter pilots, first and foremost. That flame—the burning love for flying fast jets—was rarely extinguished just because a guy pinned stars on his flight suit.

Aster smiled and nodded. "Damn straight. Life was a whoooole lot simpler, and more fun, as a fighter-jock captain." He relaxed, talking easily with Ferris and Pearson as they waited for the fighters to reach California's West Coast. They soon noted that the Eagles had joined with a waiting KC-10 air-refueling tanker. The fighters took turns topping up their fuel tanks, then headed for a spot about 200 miles to the west.

23 APRIL/OVER THE PACIFIC OCEAN/WARNING AREA W-289

"Vandy Range, Eagle flight's splitting up. We'll each squawk our own codes," Rochelle told the Vandenberg Air Force Base controller responsible for keeping all airliners and assorted surface craft far away from the swath of sky and ocean dedicated to his mission. Via the test director at Edwards, the Vandenberg controller had already cleared the two fighters into W-289, a huge "Warning Area" that normally saw rockets launched from pads and silos at the expansive West Coast air base.

"Eagle 1, I show two targets, diverging. You're both cleared for supersonic, and the sky's yours all the way to space," a voice responded. The controller grinned. *I've always wanted to say that!* He simply couldn't resist the temptation to deviate from standard FAA-comm protocol. What the hell. All communications were encrypted anyway, ensuring any eavesdroppers would only hear gibberish on the test frequency.

Rochelle would take the initial shot, timed to match the first Galileo navigation satellite's overflight. Hilton's target would appear several minutes later and in a different orbit, dictating that the fighters fly completely separate attack profiles.

"Eagle 1, you're in position. Stand by for accel in one

minute . . . stand by . . . hack!" the test conductor said, his tone professional and calm but edged with urgency. Rochelle punched the old-fashioned, large-face stopwatch Velcroed to his instrument panel. Its sweep second-hand provided an excellent rate-cue for timing his acceleration and pull to a steep climb angle. Digital numerals counting down on the head-up display directly in front of his eyes served as backup, but were less usable rate information.

Through the transparent HUD, Rochelle could see where he was headed while also monitoring flight-critical info. Bright green lines projected on the display's thick combiner glass presented an array of information pilots normally used to fly and fight without looking inside the cockpit. Today, a special set of that green symbology, tailored to this mission, would guide him through the tricky launch profile.

Rochelle verified, again, that the cockpit was configured for the pending shot. No chance to go back and do it over, so everything had better be perfect the first time. He'd flown this exact zoom-and-launch profile several times during the re-instated, highly classified test program, but had never actually fired one of the expensive ASAT missiles. Before joining the test force about a year earlier, he hadn't known the air-launched ASAT program had been resurrected, but had been ecstatic when informed he'd be leading the project.

In the wee hours of this morning, he and Hilton had rehearsed the profile several times in the simulator, too. *Ready as we'll ever be,* he thought grimly. *Just don't screw this up!* Every test pilot feared the dreaded high-profile, critical-test screw-up more than anything else, even death. He was trained, he had a good jet, and everything had been checked a dozen times. Now it was up to him to fly the profile. No easy task, but he was up to it. After all, that's why "golden arm" test pilots were tagged for this mission. Demanding, precision flying was their stock in trade.

"Eagle 1, go max power," the test controller commanded. Rochelle had already moved the throttles to military power or 100 percent. He shoved them past a detent, activating afterburners at the controller's call. He constantly cross-checked his GPS/inertial position against airspeed, accelerating quickly.

"There's the Mach, accelerating nicely," he said, knowing everybody in the test control room was watching the same parameters he was. So was that STRATCOM four-button, who'd surprised everybody by bursting into the preflight briefing, trailed by his entourage, the AFFTC commander and, of course, General Pearson, the only guy to ever do this before.

Thank God for Pearson! the pilot thought. The retired general's experience had been invaluable throughout the latest ASAT test program. But Pearson's '85 shot had been a test. This was the real deal, a no-shit combat launch to intentionally take out a satellite. Rochelle had no idea what was behind the shocking orders to destroy two European navsats, but he sure as hell didn't have time now to ponder the reasons why.

"On the profile, Roach. Keep it coming," the mission conductor said, his tone rising slightly.

"There's Mach 1.5, starting the pull . . ." Rochelle smoothly tugged the stick back toward his lap, grunting against the 3.8 gs required to point the Eagle's nose precisely 65 degrees upward and stay on the prescribed climb schedule. Angle, speed, and timing were absolutely critical. "Arming . . . and going automatic," he said. The large missile was "hot" now, its computers racing through more than 200 automatic checks. In 1985, Pearson had wryly referred to that test sequence as "a series of 210 consecutive miracles," because a single failure would have aborted the launch. The modern version of his missile tended to be less finicky, but still . . . Today, the sequence passed without a hitch and the F-15's fire control computer stood poised to command a release.

Steady, hold it . . . Rochelle focused on climb angle, airspeed/Mach number, heading, time . . . and then repeated the same parameter mantra. Time, 38,000 feet; speed, bleeding off; Mach, 1.3 . . . Suddenly, he felt a terrific thump and the F-15 leaped. Instinctively, his right arm nudged the stick forward slightly, arresting a nose upswing. The ASM-135C, a much more powerful, refined version of the ASAT missile Pearson had launched from this same area almost twenty-five years earlier, fell away. Its rocket ignited, producing a thick plume of white smoke. Rochelle finally saw the big weapon as it streaked into view from below, climbing extremely fast into the darkening sky above.

Still supersonic, the pilot eased both throttles back to military power, rolled inverted, and pulled the nose down. *Celestial Eagle II* was supposed to stay below 50,000 feet, the Air Force's limit for any pilot not equipped with an astronaut-type pressure suit, which was designed to keep one's blood from boiling in the super-thin atmosphere. Today, the fighter soared above that magic altitude before Rochelle could arrest his jet's climb rate. Nearing 65,000 feet, he was vaguely aware that his life depended on the Eagle's pressurized cockpit remaining intact for the next few seconds. He'd never been this high before.

He rolled back to wings level, nose down, and descended steadily, mentally reflying the profile. He'd done all he could. The missile was on its own. Had he hit the window? Or had he screwed the pooch?

"Eagle 1, Control. Right on profile. Couldn't have been better, Roach. Smack on time and in the window." Rochelle exhaled audibly, knowing the same relief was being expressed in the Edwards control room. But it wasn't over yet.

The ASAT missile raced higher, screaming into the thin air of Earth's upper atmosphere, then into the cold darkness of space. The missile's single-minded navigation system, making thousands of calculations a second, commanded tiny guidance rockets to fire a puff here, another there, keeping the satellite-killer on course. It would be almost a half hour before the missile and its target would rendezvous.

At that same moment, in Mid-Earth Orbit, or MEO, a lone Galileo satellite soared silently through space at roughly 17,500 MPH, tracing a circular orbit tilted 56 degrees from the equator. There would ultimately be thirty such spacecraft in MEO, broadcasting signals to receivers on the ground, in the air, and in space. Only three production platforms were now in service, thanks to repeated program delays attributed to European political bickering and erratic funding.

This particular Galileo navsat was oblivious to the F-15 far below, now turning toward California and descending, and to the long, slender missile climbing through lower-altitude orbital rings. The satellite's masters, sitting in ground stations on the other side of the world, also were oblivious to the imminent collision. The satellite sailed along in the same orbit, day in, day out, while the Earth rotated under it. Tests were conducted only when its track crossed Europe. Otherwise, it simply orbited and beamed its signals downward. At the moment, it was zinging over Asia, maintaining a constant 14,674 miles above sea level.

The ASAT killer missile continued to climb, aimed at an imaginary spot in the black, far above. It shed a burnt-out solid-fuel first-stage rocket, commanding the remaining stages to aim a Micro-Kill Vehicle (MKV) at the oncoming European Galileo satellite. An advanced staring infrared sensor finally spotted its quarry as the spacecraft appeared above the horizon. Unlike the vehicle flown in 1985, this one did not require a spin-up for stabilization. Two-plus decades of hit-to-kill, missile-defense-interceptor technology had refined the MKV to an incredibly accurate, deadly weapon.

Closing at approximately 15,000 MPH, the MKV homed on the Galileo navsat. In a flash, the two collided, scattering debris in all directions. The navsat's signals were silenced immediately, replaced by the hiss of dead radio space. It would be long minutes before the satellite's owners would discover that signals were no longer being transmitted, triggering a frantic search by radar and visual means. All they'd find were scattered pieces, most too small to detect.

Back in the atmosphere, Hilton was repeating Rochelle's precise acceleration, 3.8-g pull-up, and 65-degree climb. His ASAT, too, fired on schedule and was soon climbing skyward, aimed at another spot in space. A second Galileo satellite was headed to that same spot, its remaining life now measured in minutes.

Over the encrypted test-mission frequency, an elated test

controller at Edwards finally called. "Eagle 1 and Eagle 2, we have confirmation from Cheyenne Mountain: two for two, guys!"

Rochelle smiled behind his mask, turning in time to see Hilton slide into position off his right wing. Sierra pumped a green-gloved left fist up and down briefly. The wingman had to be grinning behind his mask, as well, Rochelle knew. *That's why they hire us. When it absolutely, positively has to be right on the money, get yourself a shit-hot test pilot!* He was jazzed. This is what he'd trained to do, the ultimate in precision flying. Today had demanded a unique blend of flight testing and combat flying rolled into one mission, one missile shot each. And they'd done it! *Frappin' awesome!*

Thirty-seven minutes later, two pale-blue Eagles in tight formation screamed above the Edwards runway in a traditional fighter overhead pattern, their centerline weapons stations empty. The first snapped into a hard left bank, then pulled away, standing on one wing as it carved 180 degrees to downwind, slowing in the turn. A five-count later, the second F-15 did the same. The two jets decelerated, their gear swung down and locked into landing position, and both smoothly curved down and back to the west, on final approach to Runway 22. Wings level and nose high, the first F-15 settled to the concrete runway, its main-gear wheels chirping and throwing a puff of burnt-tire smoke at touchdown. Rochelle held the Eagle nose-high, using its big wing as an airbrake, then gently lowered the nose wheel to the pavement, rolling toward a midfield turnoff.

Before his flight leader cleared the runway, Hilton repeated the show. Rochelle waited for his wingman, then the two jets taxied in formation across the acres of concrete that constituted Edwards's huge ramp and taxiway complex. They turned north, swinging around the tall flightline control tower, and rolled along contractor's row, the massive dry lakebed on their right. Soon, Rochelle caught Hilton's eye and pointed ahead. A sizable crowd was already waiting near the Fighter CTF's hangars.

Neither pilot would buy his own drinks tonight.

Aster, Ferris, and Pearson waited with the excited, noisy crowd, but stood to one side, by themselves. Unlike the back-slapping mechanics and chattering test team, the older, higher ranking men were silent. All three had been damned excited in the control room when a call from the Space Control Center in Cheyenne Mountain had confirmed "two kills." But now, none spoke. Behind standard Air Force–issue flight-crew sunglasses, three sets of eyes narrowed, matching the tight lines of their lips. All suffered from an odd mix of grat-ification, relief, and unease. They had carried out the orders of their president and secretary of defense, but they dreaded the fallout that was sure to come.

Watching the two F-15s turn slowly from the taxiway and amble toward the crowd, Aster leaned into Pearson's right ear, shouting to be heard. "Doug, I can't thank you enough for coming back to help us out. Your guidance and corporate memory have been absolutely critical to the success of this mission."

Pearson smiled ruefully. "My pleasure, General. It's rare that an old test pilot gets to jump back into the mix and do something worthwhile." He hesitated, staring at the F-15s for a long moment. The Eagles were only fifty feet away, starting to wheel right, toward the crowd. They obeyed the guidance of two crew chiefs marshaling them in, each chief raising one arm and motioning front to back, the other arm aimed hori-zontally to the side, directing a turn.

Pearson stuck fingers in both of his ears, then turned and shouted to Aster. "Good luck, Howard. I suspect we just un-leashed a real shit storm, and some of it's gonna fly your way." Aster glanced at the retired two-star and shot him a twisted, lopsided grin. Then he faced the two F-15s and snapped a congratulatory salute.

25 APRIL/STRATCOM COMMAND POST

Lieutenant General Dave Forester looked the part of an Army commander, dressed in creased, freshly starched battle fatigues, Aster noted as the commander entered his STRAT-COM command post. The large room's lights were dimmed, enabling its occupants to scan tabulated information scrolling on large computer screens.

Aster was bone-tired. The fast out-and-back flight to Edwards for the ASAT mission two days earlier, topping two weeks of dealing with space and global crises, had finally drained him physically, mentally, and emotionally. He knew his body and mind were wearing down, and he'd have to get some quality rest soon, but the crises weren't over yet.

Aster scanned the darkened room, taking note of the "Earthcam" displays—images beamed from cities around the world—projected on wall-size and smaller displays. During the last decade, private parties had hung inexpensive digital video cameras outside their windows or on balconies, creating a tableau of neighborhoods, busy streets, parks, and tourist attractions, from New York's Times Square to Venice Beach in Los Angeles. All fed into the ubiquitous Internet, the cameras now making up what Jill Bock called "the people's surveillance network." Since 2006, she'd urged her bosses to install "cam-monitors" throughout the wargaming center and command post so players could watch real-time events in cities targeted by their virtual bombers and nuclear-tipped missiles. Those images had made a difference. Wargamers now looked at those flesh-and-blood people going about their daily lives, and thought twice before flippantly launching even a make-believe strike during a game. The same image-driven psychology also had been introduced into the real-world STRATCOM command post recently. Aster wanted his war planners to see those faces and cities.

Today, as clouds of imminent war swirled across the globe and hung low over Aster's wargamers down the hall, dozens of men and women dressed in battle dress uniforms and flight suits hunkered over computers and moved with purpose throughout the expansive command post. From this room, Aster and his battle staff literally could run a war anywhere on Earth. Or, thanks to Jill's cam-monitors, laugh at some unsuspecting Frenchman on the Champs d'Elysées blowing his nose into his hand, then running his fingers along an elegant Citroën parked at the curb.

Forester escorted Aster to the battle cab, a raised, glass-enclosed area reserved for STRATCOM's senior staff and commander. As director of operations, Forester quickly summarized the status, and his interpretation, of events across the globe, as well as the deployment, alert levels, and overall readiness of U.S. and coalition forces. Aster absorbed the information, asking for clarification now and then.

"Copy all, Dave." He finally nodded. "But exactly where do we stand with the Iranian situation, from both political and military perspectives?" He was pointing to a large screen dedicated to the Central Command region.

"Information's still a little uncertain, sir. I'd say it's lookin' mighty encouraging, though," Forester answered. "That Shahab-4 and its nuke did *not* launch within the predicted window, and our spooks' analysts think the loss of Galileo nav information was behind Iran's 'hold' decision. NSA's picked up a lot of traffic between some Iranian rocket-forces muckety-muck and a French company that delivered those Galileo receivers to Iran a few days ago."

"And the gist of that traffic?" Aster asked.

Forester tried to stay corporate, answering the question in the tone it was asked. But, deep inside, a voice fought to burst out laughing. The Iranians were screaming they'd been ripped off by the nasty French and they wanted their money back. Forester suspected Aster wouldn't see any black humor in the situation, though. There *was* a war on, after all.

The Army general suppressed a chuckle and replied, "Mostly discussions about Galileo's current and projected status, sir. The French finally 'fessed up that two of their primary navsats are goners, and have adamantly told the Iranians to not rely on Galileo nav data for any so-called critical operations," Forester said. He smiled broadly and added, "The Iranians are royally pissed, that's for sure. They're demanding the Frogs take their damned high-accuracy, sophisticated receivers back and send a refund check, for Chrisakes!"

Aster laughed aloud, struck by the unexpected incongruence. A nuclear holocaust may have been narrowly averted, yet two idiots were bazaar-haggling over the price of a high-tech trinket, screaming at each other about a few hundred-thousand euros!

"Are we sure Iran's standing down, then?" he asked, wanting airtight confirmation.

"No way to tell for sure, sir. Intel says it appears the Guards have removed the warhead from that Shahab-4 and returned it to storage. That's based on comm traffic and some Lacrosse satellite radar data showing a convoy leaving the launch site in the middle of the night. Maybe some eyes-on info from an in-country agent, too, I suspect. We're still waiting on assessments from the latest overhead imagery, but a prelim report says there's much less personnel activity at the launch site. My take is, they're standing down, at least for now," Forester concluded.

Aster nodded, eyes still scanning the CENTCOM screen. "Let's say the Iranians are playing us, making it *look* like they've de-nuked that Shahab," he mused. "What if they *do* launch a blivet? Can we take it down?" The Galileo gambit had bought them time. But time was still a high-priced commodity.

Forester hesitated, then pointed at a smaller computer screen inside the battle cab. "We now have two locked-and-loaded missile-defense cruisers in the Med, but no Airborne Laser, as we'd hoped. It's still on the ground in California," he said. "If a Shahab launches, there's a damned good chance we'll shoot it down, because the ships are on 24-7 alert. They just can't hit a missile quite as soon as the ABL's laser could."

Aster nodded. "And I've already ordered that one and only ABL to the Far East as soon as it's back in the air. We may need it out there—and damned soon, if North Korea and China start acting up. Lee thinks that part of the world is a tinderbox and could blow at any time.

"I'm also trying to get the SecDef on the horn," the four-star continued. "I'll tell him the Galileo ASAT mission apparently did the trick. He's being pinged by the president and that dumb-shit Vandergrift almost hourly. Ol' Van's about worn out a set of worry beads, sweating over what the Euros are going to do, *if* they figure out who whacked their two navsats!"

Forester chuckled. "Sir, the SecDef's staff has been hammering us, too. You can probably save that call, though. I briefed a deputy SecDef more than an hour ago, and told him pretty much what I just covered with you. I'm sure Secretary Hurlburt's already passed that along to the president's office, as well."

Although they had defused the crisis of the hour, neither of the STRATCOM generals could predict the events now unfolding halfway around the world.

20

NEW NIGHTMARE

Hassan Rafjani leaned against his helicopter's window, looking down at devastation. A massive depression in a desert hillside was littered with twisted steel, huge chunks of concrete, flattened equipment, and odd bits of cloth that fluttered in the wind. Sand now filled what had been a very large underground complex, a virtually indestructible bunker protected by tons of Iranian sand and thick concrete-and-steel walls. The Germans who had surreptitiously built the well-camouflaged facility over several years had assured him that Parchin was impervious to all but a direct nuclear attack. Still, Parchin had been vaporized, it seemed.

From his perch a hundred feet above, he was stunned by the complete, utter destruction of what had been Iran's pride—a world-class nuclear-weapons development facility. He was numbed by the stark scene, yet anger raged within his soul.

Turning to the pilot, Rafjani jabbed a thumb downward, signaling a lower altitude. He wanted a closer look.

The pilot's eyes betrayed fear. In Farsi, the pilot inquired, "Sir, this area is highly radioactive. I fear for your safety—"

"Down, fool! Now! There is no radiation to fear! Down!" Rafjani ordered harshly.

The pilot obeyed, convinced he was committing suicide. At best, he would never father another child. Everybody knew nuclear radiation would sterilize even the most virile man. The helicopter descended until its rotor blast whipped the ground, then Rafjani raised a hand, ordering the descent halted. He drew a circle in the air, telling the pilot to slowly orbit the zone of destruction below.

Scant feet above the ground, the helicopter was permeated with the stench of burnt, rotting flesh. Yet no bodies were visible. *Death reigns here*, Rafjani thought. Indeed, no living soul moved throughout the area, but he knew hundreds of dead scientists, engineers, technicians, and production workers lay below the wind-shifted sands. A ground team of Republican Guard soldiers had arrived within hours after the Parchin facility had suddenly, inexplicably gone silent. The soldiers had not remained long, however, fearing exposure to radiation. Yes, some enriched uranium had escaped, blown free by an apparent explosion, but only a few millirems had been detected, nothing to fear if one limited his exposure time.

Initial reports indicated that an accident had destroyed Parchin. Perhaps something had failed in the uranium-enrichment process, triggering a chain reaction that had gone critical, setting off a low-grade nuclear detonation. But the man known in the West as Dagger strongly suspected something else had happened. A nuclear detonation would have set off alarms across the globe, triggering a massive outcry from every nuclear watchdog on Earth. Israel, America's Zionist proxy, would have responded immediately, launching airstrikes against Iran. And surely the Great Satan, with its space-based nuclear detonation detectors aboard myriad satellites, would have gone into a frenzy at the United Nations, blasting Iran without mercy.

But none of that had occurred. Which could only mean that the explosion had *not* been a nuclear blast; of that, he was certain. That's why he was here now. To see for himself.

Rafjani signaled to a frail figure strapped into a seat behind his own, then pointed downward. He wanted a full set of photos taken. The photographer had been sweeping the area with a video camera, but digital photos could be analyzed in more detail. While the helicopter slowly circled the zone of devastation, Rafjani tried to view the area with his "inner eye," looking for what, he wasn't sure. A pattern of debris or sand, perhaps. Something that would tell him what had caused the Parchin facility to evaporate.

His eyes were drawn to a shiny depression. Water? Grabbing a set of binoculars, he tried to focus on the glittering mass, but the aircraft's vibration made it difficult to hold the glasses steady. In exasperation, he keyed the intercom mike and directed the pilot to fly over the mass, then hover. Dagger shoved the door open, allowing him to look straight down. The shiny puddle was only a few feet under him now. Although it appeared to be liquid, a pool conforming to the sand's contours, its surface was unaffected by the rotor blast and sand that whipped up in eye-stinging clouds.

Rafjani nodded in satisfaction, slammed the helicopter's door closed and jerked his thumb upward. Blowing dust was cutting visibility to nil. Quickly, the pilot pulled up on the collective, dumped the nose over with the cyclic, and started racing forward. When he gained sufficient airspeed, the pilot pulled back on the cyclic and commenced a rapid climb-out from the swirling dust that seemingly chased the helicopter.

The hard-eyed Iranian's angular face darkened. The "puddle" had been a pool of melted aluminum that had hardened as it cooled. But there was no sign of fire nearby; no blackened timbers or other scorched material. Something had created a tremendous amount of heat in a very brief span of time, something he did not understand . . . yet.

He ordered the pilot to return to base, then stared out the window, not seeing the stark emptiness of desert flash by. He was only vaguely aware of the helicopter's distorted shadow as it raced alongside, diving and dipping across the rough terrain.

An American stealth bomber, maybe? With a new, powerful weapon that left no trace of explosive? He'd ordered the scared Guard technicians back into the Parchin devastation to measure radiation levels and search for nitrates—evidence that the site had been bombed. He'd allowed them to wear protective suits, even though he couldn't have cared less about their personal safety. They had recovered multiple samples, but a Tehran lab had insisted the nitrates only matched the high explosives used in Iran's nuclear weapons. Nothing else had been found.

It didn't matter, though. Although he had no proof, Rafjani had already informed the president and Supreme Leader that, without question, America was responsible for Parchin's destruction. And the Great Satan would be judged by Allah. The nuclear heart of Islam had been ripped from its holy underground shrine, and that evil act *would* be avenged!

Dagger swore to his God that the U.S. and its fellow crusaders would burn in a fiery Hell . . . forever.

29 APRIL/GENERAL ASTER'S OFFICE/OMAHA

"You will stop by before you leave, Admiral?" Aster asked Stanton Lee, reading the note, handwritten on Oval Office stationery. "This is very informal for the president," he remarked, as he handed it back to the retired naval officer.

"I think you are premature in your assessment of the future, Howard," Lee said. "Part of me is still hoping against hope that this is a private briefing session and no more."

"As I understand it," Aster began, still playing for a single moment of Lee's humanity to emerge, "it was you who suggested an 'old China hand.'"

"And I have the oldest hands around here. Is that what you're intimating?" Lee asked, a wry smile breaking through the stern demeanor of a prep school proctor.

Three points from half court, Aster whispered to himself, satisfied. Then he said, "I have a sneaking suspicion that this will turn into more than a briefing. There's an old saying, 'Be careful what you wish for.'"

Lee broke in. "There's an older saying, 'You get what you

pay for.' I paid for this one. I simply couldn't let the national security director get an order to send special military forces into the streets. My lord, Howard, can you imagine what that would have done?"

"I can, Admiral, I certainly can," Aster said. "In front of the entire world, we'd admit that our own civilian public safety units can no longer maintain law and order in American cities. The bad guys win." He stared hard at the old admiral. "And now you play the China card. Is it you who will keep us out of war?"

"If the president's advisors listen, we have a chance," Lee said. "But the joker in the deck is North Korea."

Almost simultaneously, Lee's and Aster's communicators went off, vibrating and buzzing like two angry bees. The admiral and the general locked eyes across the expanse of Aster's office, each feeling the tug of his respective responsibilities drawing him away from what he sensed was an unfinished conversation. There was much more to say, but each man stepped aside to pick up his call. In Tehran, the Iranians wondered how Parchin had exploded. In Beijing, the Chinese questioned why an American space vehicle had shadowed their nanosat array. In Omaha, Admiral Lee's and General Aster's questions, like so many in time of war, were left hanging in the air like morning fog.

4 MAY/SIDEWALK CAFE/TEHRAN, IRAN

The Tehran evening hinted that cooler air would soon replace the stifling heat of an early-summer day, prompting COBI, the senior in-country CIA agent, to ask for an outside table. He had intentionally arrived early to beat the crowd of diners that usually filled the café's tables quickly.

COBI placed his order with a surly waiter, then surveyed the area. He was well positioned on one side of the small space reserved for patio dining. One of the marks he was to observe this evening would be Russian, a person almost assuredly from the cool of Moscow. That led the agent to predict his targets would meet out here rather than deal with the hot, crowded, noisy café's interior.

He mentally reviewed what little information Ari, his longtime Iranian contact, had relayed, delivered with uncharacteristic urgency and concern, only two days ago: A Russian security agent was inbound; no description, but a person noted for his specialized technical expertise. The Russian was to meet a hard-line Iranian agent here, tonight. The latter was known to be a powerful, shadowy figure close to Iran's leadership—and involved in the nation's nuclear program, as well.

Based on voiceprints made by the National Security Agency from intercepted cell phone communications, this was the same Iranian who had financed that devastating maser strike on the International Space Station. COBI's teeth clenched unconsciously as he thought about the Iranian agent. *One bad-assed dude,* he concluded. The covert CIA operative would like nothing better than to put two silenced 9-mm rounds behind the guy's ear. It was *that* S.O.B. who'd personally ordered the death of those five space station astronauts, leaving them and a multibillion-dollar facility drifting lifelessly in space.

It had taken some serious work, and a lot of American greenbacks liberally spread around Tehran's underworld, but COBI had managed to get a photo of the Iranian agent whom the Russian would be meeting tonight. Known only by his CIA-assigned code name, "Dagger," the guy had deep and deadly roots in Iran's radical Islam community. The blue-eyed, hawk-faced man had proven to be one of the Ayatollah's most devout, loyal servants over the decades, steadily rising to a position of considerable power within the tight circle of clerics that now controlled Iran's political and military cabal. But that was the extent of information in the CIA's "Dagger" file. *Not much to go on,* the agent concluded silently.

COBI opened a laptop computer, appearing to be simply another scruffy university student surfing the Internet via wireless connections offered by this particular café. He was momentarily struck by the incongruity of simply activating his modern computer in Tehran and tapping into a high-speed wireless connection. Although conservative mullahs had steadily tightened their grip on Iran after Mahmoud Ahmadinejad's election in 2005, ousted reformers had managed to accelerate the spread of high-tech communications throughout the nation. Why the mullahs allowed it, nobody on the outside was certain. Perhaps a quid pro quo for the masses swallowing myriad regressive policies the mullahs had imposed on Iranian society. COBI didn't know and didn't particularly care. His job was to acquire information inside Iran and send it to CIA headquarters. Smart analysts back at Langley pored over that information and decided what it meant.

COBI felt the Iranian agent's cold presence before he saw him, glancing up to see the tall, hawk-faced man with deeply set, uncharacteristically blue eyes carefully perusing the outdoor seating area before selecting a table. The newcomer placed a thick manila envelope on the table's surface as he seated himself, but left a hand resting on the packet. The CIA agent had guessed well. The Iranian and Russian would be seated in exactly the right spot for easy surveillance, at a table somewhat removed from the bustle of waiters and arriving diners. COBI adjusted the small-screen computer slightly, pressed a couple of keys, and noted a nondescript icon appeared in the toolbar near the display's top edge. He was ready, his high-tech surveillance system on standby.

Long minutes crept by, but the Iranian agent seemed unperturbed. He nibbled on a small plate of dates and cheese. Two glasses of water sat on the table, indicating that he would be joined at some point by another. COBI ignored the agent, appearing to focus intently on the computer's screen, oblivious of his surroundings. His fingers raced furiously across the keyboard, accentuating his seeming intensity. Dagger had perused COBI closely for a short period, then apparently decided he was of little interest. Just another driven, unkempt student, he'd apparently concluded.

A thick-necked man in rumpled slacks and an equally

wrinkled open-collared shirt finally appeared at the Iranian's table, prompting Dagger to stand in greeting. *There's my Russian,* COBI thought, satisfied. The two shook hands formally, then each took a chair, facing the other. The Russian carefully placed a bulky aluminum-sided box under the table, inside his left foot.

COBI glanced around him, appearing to reassess the ambient lighting as darkness approached, then adjusted the laptop computer's hinged, integral screen. In fact, he was carefully aligning an invisible laser beam, ensuring it crossed the roughly seven feet that separated his table from that of the Iranian and Russian. A beam shot from the top edge of the computer's display and bounced off a water glass directly in front of the Russian. The reflected laser energy was detected by a vertical row of nanosensors buried along the screen's tilted edge. When everything was lined up properly, the small icon pulsated slowly.

COBI leaned back, yawned widely and scratched his thick beard, then his ear. That casual motion powered up another microscopic system, a tiny earpiece nestled in his right ear. He nodded ever so slightly, satisfied that the covert laser-based listening system was performing perfectly. That invisible laser beam bounced off the Russian's still-untouched water glass, picking up ever so tiny vibrations, and returned the now-modulated laser signal to the row of nanodetectors. Lightning-speed processing by a special microchip inside the computer turned that modulated laser signal into coded pulses. Those, in turn, were transmitted via covert ultrawideband signals to the tiny wireless receiver/earpiece inside COBI's ear. His wild, curly hair and beard covered the device.

A few taps on the computer's keyboard refined the audio in his ear, ensuring crystal-clear reception. The water glass was acting as a very effective microphone, picking up every word spoken by the targets.

The two men were hunched over the small table, talking in low voices. The Russian appeared uneasy, glancing about as if concerned he'd be overheard. Dagger's hard eyes stayed locked on his tablemate, listening intently. As COBI had hoped, the two men were conversing in English. That made his job considerably easier. He was fluent in Farsi, but had bet that the Russian wouldn't be. English would be the compromise language, he'd hoped.

". . . sophisticated system. Much better than the clumsy American GPS and much more reliable than the European's immature Galileo system. You know Galileo is much degraded now, yes?" the Russian asked importantly. The Iranian simply nodded. Responding to the surly Iranian waiter who appeared, the Russian broke his dialogue long enough to place an order. He hadn't touched his water glass, just as COBI had anticipated. Something stronger than water was the usual Russian preference.

"How accurate is your Glonass system?" Dagger asked pointedly. "Accuracy is very important to my people."

The Russian flicked a hand dismissively. "One meter, maybe two meters at . . . let's say, ranges of interest." He again glanced furtively about. *The new crop aren't well-trained or seasoned,* COBI concluded. Not like the old KGB. These new Russian agents, if that's what this guy was, were much more clumsy than their predecessors. *Maybe he's just a geek looking to make a fast buck,* COBI thought, then shrugged. Again, it didn't matter.

The Russian pulled a folded sheet of paper from his hip pocket and spread it on the table. Conversation became even more muted as he pointed to the sheet, explaining a number of technical details. Dagger leaned forward, soaking up every word, it seemed.

"I understand," the Iranian finally said, taking the sheet and slipping it inside a coat pocket. "Your Glonass item is suitable for my needs. And your compensation is as we agreed."

The Russian again glanced around. Eyeing the pair with well-trained peripheral vision, COBI thought he saw the Russian's tongue flick across now-dry lips. "You must appreciate how difficult it was to bring such, ahh . . . advanced technology into your country," he stammered. "Much more difficult than you indicated it . . ."

Dagger's right hand shot across the table and grabbed the

Russian by his wrist. A pained gasp escaped the visitor, prompting an involuntary attempt to jerk his hand free. It failed.

"The arrangement made last week stands. The price was for delivery of a properly functioning device. You are delivering it now, I presume. But we must first check it, see that it performs as you claim. Half your money is here—one hundred seventy thousand euros, just as you agreed," he said, his quiet, accented English laced with venom. "The other half you will receive tomorrow, *if* our tests confirm its performance meets our needs. Do you understand, my friend?" Dagger asked firmly.

The Russian grimaced and nodded, his crushed wrist firing white-hot spasms of pain up his arm. The Iranian eased his grip on the other man, leaned back slowly, and slid the thin folder closer to the Russian with the other hand.

Clearing his throat, the Russian smiled nervously, his head continuing to bob. Trying to regain lost composure, he lowered his voice, adding, "And your Shahab . . . will it be for the American base again? Or is it Tel Aviv . . . ?"

"Silence, you fool!" Dagger hissed, jerking forward again. The Russian retracted both hands quickly, pulling the envelope to his chest in the same motion. "You know *nothing!* Now go! Do not leave the hotel until you hear from me, or you die!" In one surprisingly rapid, fluid motion, the angular Iranian swept the aluminum-sided case from beneath the table, straightened and quickly exited the café area. The Russian stammered something unintelligible, turning and raising a hand as if to halt the other man, but Dagger was gone, striding quickly along the crowded sidewalk.

The surly waiter practically ran to the now-half-vacated table, speaking rapidly in Farsi. The Russian disgustedly handed him a colored paper bill, his hand shaking visibly. Dagger had stiffed him for the tab, COBI noted, fighting to stifle a grin.

The CIA agent focused intently on the computer screen, apparently reading its contents with great interest. But his mind was racing, re-creating events he'd just witnessed.

There could be no doubt as to what had transpired. The Iranian had purchased yet another satellite-navigation receiver. And not the garden-variety type one might buy at a department or sporting goods store in the States.

The bastards have a Russian Glonass satnav receiver! A sophisticated system as good as any military GPS unit, well suited to guiding a Shahab-4 and its payload to any target within a several-hundred-mile radius! Despite years of experience in the volatile, unpredictable Mideastern and Persian region, COBI was stunned to the core of his being. Iran was back in the nuclear missile business—already! And just as North Korea and China were sabre-rattling in the Pacific. *Christamighty!*

Although his CIA colleagues and Washington's political and military leaders were still congratulating themselves for defusing a very complex global crisis that had come dangerously close to unleashing a nuclear fire-and-brimstone holocaust, this information would change all that. Iran was still determined to launch a nuclear attack. When and against whom, he didn't know. But COBI was absolutely certain of one thing, a certainty that chilled his soul: Another horrible nightmare had begun.

SPACE
WARS

WAS JUST THE BEGINNING

In *Space Wars,* Michael J. Coumatos, former U.S. Space
Command director of wargaming, and William B. Scott, Rocky
Mountain Bureau Chief of *Aviation Week & Space Technology,*
joined forces to depict how the first hours of World War III
could play out in the year 2010. Now, the story continues in…

COUNTERSPACE

MICHAEL J. COUMATOS,
WILLIAM B. SCOTT, AND WILLIAM J. BIRNES

....

"Former navy flier and wargamer Coumatos joins forces with
former air force aviation engineer William B. Scott and
lawyer William J. Birnes…to produce this engrossing
piece of military futurism…. Presages real possibilities."

—*Booklist* on *Space Wars*

978-0-7653-2232-6 • 0-7653-2232-3
IN HARDCOVER IN 2010

www.tor-forge.com